T0076170

ARTIFICIAL INTELLIGENCE

www.royalcollins.com

ARTIFICIAL INTELLIGENCE

RISKS, REWARDS, AND THE FUTURE

KEVIN CHEN

ARTIFICIAL INTELLIGENCE:
RISKS, REWARDS, AND THE FUTURE

KEVIN CHEN

First published in 2023 by Royal Collins Publishing Group Inc.
Groupe Publication Royal Collins Inc.
550-555 boul. René-Lévesque O Montréal (Québec) H2Z1B1 Canada

ISBN: 978-1-4878-0997-3

To find out more about our publications, please visit www.royalcollins.com.

Contents

CHAPTER 1

The Golden Age of Human Intelligence

1.1 The Ups and Downs of Artificial Intelligence

It's probably not news to you anymore that the term artificial intelligence (AI) has been brought up more and more frequently around us. With people's attention being increasingly drawn to the most recent technological updates and research outcomes in this field published online and in books, AI is no longer an academic terminology that exists only in labs but has become a hotspot of popular science in the Internet age.

Has it occurred to you that changes brought by human intelligence have already emerged here and there in our everyday lives? The news programs you turn on are chosen by an AI algorithm. The front pages of your shopping websites are covered by items that you're most likely to be interested and purchase—also the result of AI calculation. The technologies that make all these possible, which are no less inferior to that which made AlphaGo defeat its human competitors, are bringing unprecedented convenience to people's lives in many more aspects.

At the same time, it should be noted that AI as a sub-discipline in computer science has merely come into being in the last 70 years. During this time, generations have witnessed its adventurous journey filled with confusion derived from unsolved puzzles and debates among different schools. The ups and downs of AI development have left people wondering, where does the current path of this study lead to? In order to answer this question, we need to look back in history and see where this path came from.

• From Ancient Fantasy to Modern Reality

The thinking and longing for man-made mechanisms with intelligence began long before the term artificial intelligence was invented.

It is not a coincidence that ancient tales and legends of exceptionally skilled craftsmen building artificial figures and bestowing them consciousness and senses are found all over the world. In Greek mythology, there are Hephaestus' golden robots and Pygmalion's ivory sculpture that came to life. In Chinese classic *Liezi*, the "Tangwen" chapter records the story of master Yan who built a robot that looked exactly like a real person. In Jewish legends, there are earth figures displaying certain features of life. In Indian myths, there are robot guards built for the purpose of protecting Śarīra, relics of Buddha. (mimicking mechanisms in the form of human figures built by ancient Greeks and Romans)

The first walking robot recorded in legendary history was Talos, the gigantic, animated bronze warrior built by Hephaestus, the Greek god of metalworking, around 2,500 years ago in Crete. According to *The Iliad*, Talos was in charge of guarding the Crete islands in the Trojan War. At the feast of gods, his food was served in animated mechanical tripods.

In her book *Gods and Robots*, Adrienne Mayor refers to Alexandria as the primeval Silicon Valley, for it was the birthplace of countless robots.

These examples of ancient "robots" were, of course, not relatable to the modern concept of AI in any sense. But the animated mechanics in the forms of human beings displayed people's early attempts to replicate themselves.

The influence of such lasting endeavors can still be seen even in the present era, the vision of artificial general intelligence (AGI)—machineries with super intelligence that allows them the abilities to learn, understand, and perform any intellectual task like a human being—which sets the world in an entangled sentiment of fascination and concern. Nevertheless, Jean-Gabriel Ganascia, who is a professor of computer science at Sorbonne University, points out that based on our current level of technological development, discussions of man-shaped objects endowed with life and AGI are still dealing with mere fantasies instead of scientific facts.

The truth is, the topic of artificial intelligence has always been dominated by mere fantasies until the 1940s, when studies in calculating and communicating became the main focuses in world science due to the excessive needs of these technologies by the belligerent states in WWII. With the invention of information science and the computer, scientists began to seriously consider the possibility of creating artificial intelligence grounded in scientific reality.

In spring 1935, 23-year-old Alan Turing, who was appointed to a fellowship in King's College, the University of Cambridge, came across the 10th problem in Hilbert's problems (23 problems in mathematics published by German mathematician David Hilbert in 1900) for the first time: "Find an algorithm to determine whether a given polynomial Diophantine equation with integer coefficients has an integer solution."

Turing correctly perceived that the key to solve this problem lay in the clear definition of "algorithm." The original definition of the term given by Hilbert was probably "a calculation that can be finished according to certain, limited steps without innovations from the calculator," which can be considered both accurate and creative at a time when computer was not yet invented. But Turing's idea was more straightforward, and he defined "algorithm" as "calculation that can be completed by a single piece of machinery."

In 1936, Turing published the seminal paper "On Computable Numbers, with an Application to the Entscheidungsproblem" in an authoritative journal of London (*Proceedings of the London Mathematical Society*) which introduced the epoch-making Turing machine to the field.

In 1950, in his other paper "Computing Machinery and Intelligence," Turing introduced for the first time the assessing principles for artificial intelligence, which was since known by the world as the Turing Test. The test should be carried out between a tester (a human) and a testee (a machine) by having the tester proposing random questions at the testee through a specific set of instruments. If over 30% of the testers cannot distinguish which answer is given by a human counterpart and which by the machine after communicating for five minutes, this machine can thus be considered passing the test and possessing artificial intelligence.

Basically, the Turing Test redefined artificial intelligence with a behaviorist approach and paralleled it with the intellectual performances of the algorithm. One major weakness about this test that has received much criticism is the controversial definition of "intelligence" being "the ability to mimic human behaviors," for it fails to assess the internal mental process that involves the unique features of the human brain such as decision-making and creativity that are critical in bringing forth any intellectual performance.

Despite its drawbacks in design, the Turing Test was a groundbreaking attempt in the development of artificial intelligence, especially in providing it with the necessary theoretical foundation of becoming a subject in science, which then provoked endless thinking over human beings' existence within the universe and in the future.

In August 1956, a group of scientists gathered in Dartmouth College, US, to attend a workshop on a quite abstruse, almost unearthly topic, the simulation of human intelligence in learning and in other aspects of machines. Among the participants were John McCarthy, founder of the LISP language; Marvin Minsky, expert in artificial intelligence and cognitive linguistic; Claude Shannon, the "father of information theory;" Allen Newell, researcher in computer science and cognitive psychology; Herbert Simon, winner of the Nobel Prize in economics; etc. During the approximately six to eight weeks that the workshop lasted, the members brainstormed on various topics including automatic computation, programming language, neural networks, theory of computation, self-improvement (on robot learning), abstraction, and creativity. Although they didn't come to consensus on every topic, the term "artificial intelligence" was coined and finalized throughout their discussions.

This brainstorming session was later referred to as the Dartmouth Summer Research Project on Artificial Intelligence and was generally considered the founding event of AI as a field.

• Up and Down, Up and Down

The several years after the Dartmouth Workshop can be called the era of discovery. For the majority back then, many of the programs developed at this time were amazing—magical, even—for it exceeded their imagination that machines could be so "intelligent" that they could be used in solving algebra problems, proving geometry theories, and learning English.

In 1961, the world's first industrial robot Unimate was put to work on a General Motors assembly line at the Inland Fisher Guide Plant in Ewing Township, New Jersey. Five years later, in 1966, Shakey, the first general-purpose mobile robot (and the robot who smoked) was born. Another robot which also came out that year was Eliza. "She" was programmed to carry out conversations that resembled psychological counselling with people according to a pre-implanted DOCTOR script. Eliza could be viewed as the grandma of Alex and Siri—a primary staged virtual assistant without a human form or a voice.

At the time when Eliza was born, tendencies in experimenting problem-solving with machines and voice-interpreting languages were already emerging. But enabling human intelligence such as abstract thinking, self-recognition, and natural language processing in machines was still an unreachable future.

Nevertheless, these uncertainties did not impede researchers' ambitions and enthusiasm in exploring the field of AI. Scientists were highly positive that the world would see fully intelligent robots in 20 years. All research regarding artificial intelligence was receiving unconditional support. Joseph C. R. Licklider, who was then director at the U.S. Department of Defense Advanced Research Projects Agency (ARPA), believed that his team should provide subsidies to the scientists rather than the projects, and he encouraged his researchers to pursue whatever topic that interested them.

But their optimism would soon be crashed in the first winter that was about to fall upon the field.

At the beginning of the 70s, AI received much criticism for being nothing more than "toys," for even the most advanced AI program could only tackle the easiest part of the problems they were designed to solve. The obstacle seemed insurmountable.

To make matters worse, the researchers soon ran into money problems because of their false and overly optimistic prediction on the challenge they were to face. In 1973, mathematician Sir James Lighthill commented on AI's achievement of its "grand objective" as a total failure, and this brought the AI studies in the UK to its nadir. As scientists failed to present the results as promised, their sponsors (e.g., the British government, the Defense Advanced Research Projects Agency, and the National Research Council) gradually reduced or withdrew the funding, especially for those projects without a clear focus. For example, NRC ceased its monetary support in AI after giving US$20 million. DARPA reported being heavily disappointed by CMU's research project on voice recognition and cancelled its funding of US$3 million per year. In 1974, it was difficult to find sponsorship on AI projects anymore.

Fortunately, new possibilities began to show up in the 80s with an AI program called Expert System that could solve problems in a specific area based on the patterns it inferred from a list of relevant information through deductive reasoning. The Expert System gained worldwide attention and was applied widely in companies.

The development of the Expert System can be traced to the software program Dendral, which was designed in 1965 and could distinguish different compounds by reading the spectroscope. In 1972, the backward-chaining expert system MYCIN could diagnose blood clotting diseases with a success rate of 69% (while the success rate of medical specialists was 80%). In 1978, XCON, used for automatic configuration of components for customers during the process of selling computers was born. It was the first expert system ever applied for commercial purposes and was welcomed excitedly by the market.

The economic success aroused social interest and support in AI research again. In 1981, the Ministry of Economy, Trade and Industry (METI) of Japan allocated US$850 million for the fifth-generation computer project with the goal of developing machines with the abilities to interpret languages, illustrate diagrams, converse with people, and reason like humans. Other countries quickly responded to this proposal. In 1984, the Alvey project costing US$350 million was initiated in the UK, and Microelectronics and Computer Technology Corporation (MCC), founded by a US company association, was ready to sponsor similar mass-scale projects in AI and information technology. DARPA also organized the Strategic Computing Initiative, and its investment in AI projects in 1988 was three times of that in 1984.

However, history always repeats itself in surprising ways. This decade of revitalization was followed immediately by another winter that brought great depression in the field.

The term "AI winter" was invented by researchers who survived the funding cuts in 1974. They noticed the public's extreme enthusiasm in Expert System and predicted that soon this frenzy would turn into disappointment. Unfortunately, their prediction was proved correct. The AI industry once again experienced a series of financial crises from the end of the 80s to the beginning of the 90s. It started with a sudden decline in the market demand in AI hardware. The computers developed by Apple and IBM had been constantly improving over the years and have exceeded the expensive Lisp machines made by Symbolics and other companies in function by the year 1987. The old products lost their places in the market and an industry worth US$500 million collapsed overnight.

• The Rise of the AI Era

There are many theories why the original dream of the AI industry—attaining human-level intelligence—which captivated the world's imagination in the 60s, never came true. In the end, AI has been divided into several sub-fields mutually independent of each other. Sometimes, new terms are created to avoid "artificial intelligence," which has been heavily overused.

Now, in its mid-50s, the AI industry has finally accomplished some of its primary goals, and it has developed in an ever more cautious manner to be ever more successful. It is undeniable that AI has outrun human capabilities in many subjects, such as in playing *go* or Texas hold'em, proving math theories, sorting through mass data and constructing knowledge, recognizing voices, faces, and fingerprints, operating vehicles, processing huge numbers of documents, and automatically operating logistics and manufactures. It can now even recognize and mimic human emotions and play the role of care givers or nursing attendants. Basically, AI has permeated every aspect of our lives.

In the present era, deep learning and reinforcement learning have become the dominant focus in the field. In 2007, Fei-Fei Li, a Chinese-American AI scientist, and her team built a large visual database called ImageNet for visual object recognition software research. It propelled deep learning and big data to the forefront and attracted significant funding. A popular view is that the ImageNet annual challenge in 2012 has started a new wave of the AI revolution.

In the last 10 years, AI wrote news articles, contended for an exclusive, and learned to recognize cats through big data training. IBM's Watson computer won first place over two champions in *Jeopardy!*, Google's AlphaGo defeated the world's best professional *go* players, and Boston Dynamics' Atlas learned how to do a triple jump. During the COVID-19 Pandemic in 2020, AI contributed to medical services by accurately detecting individuals with high temperatures, having drones carry out patrol duties and broadcast orders for the police, and assisting in the analysis of CT images.

These achievements are based on the combined effect of more powerful chips, the popularization of Cloud Services, and cheaper hardware.

The application of AI on a large scale is also made possible thanks to mass training data and highly efficient calculation offered by Graphics Processing Units (GPU). It takes considerably less amount of time and infrastructure in the data center when the GPU is applied in training Deep Neural Networks. In addition, it is also used for operating machine-training models with much bigger data size and container volume at a lower cost. Compared to using Central Processing Units (CPU) alone, using GPU, which has thousands of computing cores, allows for attaining 10 to 100 times of application throughput.

At the same time, the price and size of AI chips are constantly becoming lower and smaller. In comparison to that in 2014, the price of chips in the world market in 2020 has been reduced by around 70%. With the continual development of big data technology, we also see a reduction in the acquisition cost of marking data that is essential in AI learning as well as a major increase in its data processing speed. Moreover, the ongoing iteration of the Internet of Things (IoT) and telecommunications technology provides fundamental support for AI development in terms of infrastructure. In the year 2020, the number of equipment connected to the IoT reached 50 billion. The 5G networks, which is a milestone in telecommunication history, are bringing the fastest information processing speed for AI of 1 Gbps.

All these changes indicate the rising of a new era—the era of AI.

1.2 The Flourishing of AI Demand

The fundamental aim of the speedy implementation of AI policy is to form a flourishing AI market, but the process also depends on the conditions of the existing market demand and supply. Currently, the market displays high demand for AI technology from all terminals: C- customer, B- business, and G- government.

• C-Terminal Demand

The C-terminal demand for AI represents ordinary people's concerns about issues such as health care, entertainment, and travel that are closely related to daily lives. As people's living standards improve along with social development, they find that AI can perfectly satisfy their expectations for an intelligent life.

Right now, population aging has become an ever more serious problem in China and the need for an intelligent upgrade is desperate. Ever since China stepped into an aging society by the end of the last century, the population of those 60 years old and above increased from 126 million to 249 million between the years 2000 and 2018. This made the percentage of old people in the Chinese society increased from 10.2% to 17.9% over 18 years, which was over twice the average speed in the world. Moreover, this momentum will still be continued for a relatively long period of time, which means it will bring challenges that impede economic growth and social development, such as shortage in labor resources and increase in labor costs.

In *China's Medium- and Long-Term Plan for Proactive Response to Population Aging* released by the State Council of China at the end of 2019, it is explicitly emphasized that scientific and technological innovations should be fully exploited as the leading force and strategic support in positively confronting the problems of population aging. Apparently, in the face of the prospect of losing workforce and the negative influences brought by population aging, supplementing and improving the laboring population with AI and robots as well as upgrading industries through intellectual approaches are both inevitable and necessary.

In the fields of education and medicine, industries regarding people's livelihood, AI is also widely used and is undergoing constant improvements that aim to build a comprehensive innovative intellectual service system. In medical services, AI plays an important role in improving medical levels as well as in more specific areas such as monitoring pandemics, diagnosing diseases, and developing drugs as seen during the COVID-19 pandemic. In education, the application of AI quickens the speed in developing an open and flexible education system that respects individualities of students, which then helps to enhance teaching quality and promote education equality.

• B-Terminal Demand

The application setting and demand for AI in business are relatively more specific. For example, businesses have a robust demand for increasing efficiency, which is exactly what AI is good at. Therefore, AI is quickly adopted in all fields.

The digital economy, represented by the Internet, has been the most prominent component in China's rapid economic development for the past 20 years. According to a report given by China Internet Network Information Center (CNNIC), by March 2020, the population of netizens in China had reached 904 million and the Internet penetration rate 64.5%. It was the 13th year since China had possessed the largest Internet subscriber community in the world, and every basic necessity of life in the country, clothing, food, housing, and transportation is going online. The proportion of digital economy in GDP 2018, estimated by Tencent Research Institute, in its "Digital China Index Report, 2019," was 33.22%, a vital force in contributing to the expanding economy of China. However, in recent years, new challenges began to emerge. With the popularization of the Internet, the Chinese netizen population growth has slowed down in front of an enormous base number. The advantage of a large population is gradually disappearing, and the needs for structural transformation and quality upgrade are urgent.

The "computation revolution" led by new technologies such as AI, big data, and cloud computing is proposing a new outlook for the Internet after its previous innovative transformations. On the one hand, with broad accessibility to the Internet, these technologies ensure further decreases in cost, better connections, and higher service quality and efficiency in various connecting lines. In addition, they can create new applications and service scenarios, which more effectively address the individual needs in the market and enlarge the space for a high-quality growth of the economy. On the other hand, the scale development of these new technologies will also form new high-tech industrial ecologies that appeal to capital and talent resources. It will promote economic restructuring towards areas of high-quality development and eventually complete the upgrades in an intellectual economy and society.

• G-Terminal Demand

Building a Digital Government or an "E-government" is imperative. The first stage is to establish informatization in vertical business management with a focus on facilitating and saving costs for users. The system is still government-centered, and the technology emphasizes a service-oriented structure. It means that government offers services online in a passive mode.

The vertical business management informatization can be viewed as the "electronic government" stage. It will be mainly led by the government IT department and executed by its technical team. Its assessment standard will be the proportions of online services, meaning the frequencies of services offered via mobile devices, integrated services, and the application of e-channels.

The second stage in building a Digital Government is to construct an open government. In this stage, the governmental services will change to an active mode. The digital system will become citizen-oriented presented by a more accomplished web portal for users. The system will be accessible to the external society that can benefit from open data, and it aims for Co-creation Services.

The technological focus in this stage will switch to the API-driven structures. Leadership will derive from data-driven, and emphasis will be placed on API development and management to support big data. Its main assessment standard will be the quantity of the open database and that of the apps created based on open data.

When AI is applied according to the governmental needs in addressing livelihood issues of the people and in improving working efficiency, it can provide renewed impetus to government construction and city development.

With the help of AI, the establishment of an intellectual Digital Government is a future that can be anticipated. It is a future that government applies open data and AI to innovate its service mode through a digital approach. Its services will be foreseeable and can be predicted by various contact points, allowing more frequent interactions between the government and the citizens based on the former's increased ability in anticipating the people's needs and preventing emergencies.

To summarize, the C-terminal demands are complex with a general emphasis on product and experience. B-terminal and G-terminal demands are more specific and focused with an emphasis on efficiency. But the demands from all three terminals are robust at an equal level, which provides an important motivation for AI advancement in the long run.

1.3 Digital Revolution Driven by AI

In human history, the transformation of hunter-gatherer societies to agricultural societies was marked by the revolutionary acquisition of farming skills. 200 years ago, steam engines replaced animals in production in the First Industrial Revolution, which demonstrated the beginning of global industrialization. Before long, the viability of electricity as a source of energy triggered the Second Industrial Revolution.

These critical moments in our history witnessed the changing of human civilizations, as generations after generations continued to establish new models of perceiving and redefining the world around them each time when they made a breakthrough in productive powers, which brought new means, new bodies, and new factors of production to the table. As objective material forces created by and then applied in the interacting processes change to human beings and Nature, productive powers epitomize the standard of development of an era.

• Releasing the Digital Productive Powers

The term "productive powers" was first coined by François Quesnay, the founder of the Physiocratic school in France, to emphasize the value of land and population to a nation's wealth accumulation. The Scottish economist Adam Smith, who was heavily influenced by this theory, believed that productive powers equaled productivity, and its source for improvement lay in the continued refinement of social division of labor. British economist David Ricardo, on the other hand, considered productive powers to be different sources of "natural powers," such as capital, land, and labor.

The first basic framework on productive powers was introduced by German economist Friedrich List in 1841. Karl Marx completed it into a theoretical system, and he illustrated in his *Das Kapital (Capital: A Critique of Political Economy)* the three components of production, human labor, means of labor, and subject of labor.

Human labor refers to people with certain labor skills and labor experiences who participate in social production by contributing their physical or mental powers. Means of labor refers to object or material conditions (mainly represented by production tools and instruments) that laborers applied to subject of labor. Subject of labor refers to the things, including resources obtained directly from natural environment and processed raw materials, that are reformed during the production.

Human labor is the most active component among the productive powers. In different stages of human civilizations, the features of the laborer's production activities, the composition of the laborer communities, and the relationships between men and nature varied significantly.

In agricultural societies, people's survival depends on limited excavation of the land resources through heavy physical labor. In industrial societies, much of the burden of physical labor is taken over by machines. Now, amid the Digital Revolution with the mass adoption of intellectual instruments, people's ability in perceiving and interacting with the world has been brought to an unprecedented level. Not only did excessive physical labor become unnecessary for people, but digital productivity has also replaced us in dealing with much repetitive mental labor. That means we are using much less time to create much more material wealth.

Together, the means of labor and subject of labor are also referred to as means of production. It can only produce effects when combined with human labor, because it is people's labor activities that initiate, adapt, and control the material exchange process between men and nature. The production tools and instruments are of special importance in the system because they reflect the depth and width of human interaction with nature and are the objective embodiment of social productivity and economic development.

Regarding the means of production, Marx pointed out that the distinction between different economic times does not lie in what they produce but in how and by what means of labor. Therefore, means of labor, which is mainly displayed in the forms of production tools and instruments, is also one of the fundamental standards in classifying social patterns.

We have traveled a long way from the time of slash-and-burn agriculture to that where we used iron-and-cattle-plowing, and eventually to the present age when machines are replacing human labor. In 2017, Chairman Xi Jinping claimed that "Digital is a new factor of production, the fundamental and strategic resource, and essential productive power in the time of the Internet economy." Within the context of the digital era, the three components of productive powers (human labor, means of labor, subject of labor) are simultaneously facing immense transformations.

CHAPTER 2

A Map of AI Technology

2.1 Machine Learning

• Machine Learning is an Imitation of Human Learning

Machine learning is the most influential technology in the booming developments in AI. As a subdivision in AI technology, machine learning is also a way of bringing AI into reality. Broadly speaking, machine learning refers to endowing machines the ability to learn so that they can be used in tasks that cannot otherwise be accomplished by direct programming. In a practical sense, machine learning indicates the method of predicting using a data-based model.

This concept of machine learning was already introduced by Alan Turing in his paper on the Turing Test as early as 1950.

In 1952, Arthur Samuel from IBM, who was praised as "Father of Machine Learning," designed a program that could learn how to play checkers. It could improve its own skills in the game by observing the movements of the checker pieces and constructing new models accordingly. After playing with the program for many rounds, Samuel noticed that it had become an increasingly qualified "player" as time went by. His experiment with the program thus overthrew the conventional understanding that "machines will never surpass humans in programing and learning," and he came up with the term "Machine Learning" in 1956 for the first time. He conceptualized it as "An area of study of endowing computers the ability to learn without directly programming on the subject matter."

Another definition of Machine Learning given by the "The Godfather of Global Machine Learning" Tom Mitchell is as follows. With a certain task T and performance measurement P, if a computer program can achieve self-improvement in T under P following experience E, we consider this computer program learning from E.

As time changes, the connotation and range of Machine Learning (ML) keeps evolving as well. Generally, it is agreed that the processing system and algorithm of ML are recognition models for making predictions based on implicit patterns in data, an important sub-principle in AI. In addition, ML is a multidisciplinary subject that relates to the study of probability theory, statistics, approximation theory, convex analysis, computational complex theory, and many others.

Research in Machine Learning focuses specifically on how computers imitate or realize human learning behaviors to acquire new knowledge and skills, as well as to improve current performance by reconstructing an existing knowledge foundation. ML obtains knowledge and patterns from sample data and then applies it in making practical inferences and decisions. A major difference between ML and regular programs is the necessity of sample data, which implies a data-driven method.

We may comprehend ML as similar to the growing and learning process of human babies. As we know, most human intelligence is learned instead of being born with it. In order to build up the cognitive ability that we do not own at birth, infants need to be continuously exposed to information from their surroundings to receive external stimuli that can activate their brains.

For example, when we teach children abstract concepts such as "apple" and "banana," we will need to repeatedly mention the word while showing their connections with real apples and bananas. The children will need long-term training to construct an understanding of the abstract concepts of "apple" and "banana" and to apply their understanding to the real world they perceive.

As humans live and grow as a species, they accumulate numerous experiences that gradually become their history. By summarizing these experiences from time to time, people understand patterns of living, which they draw upon to make predictions of the future and unknown problems to guide their current lives and works.

ML is of the same principle. For example, when we "teach" AI programs the ability to recognize images, we will first need to collect huge numbers of sample images and mark out their categories such as "bananas," "apples," or "others." The algorithm will then learn (be trained with) these sample images and form a model, which is knowledge it has learned from summarizing the samples. Next, it will be able to use this model in identifying new images.

The processes of "training" and "predicting" in ML can correspond to that of "summarizing" and "inferencing" in human learning. Therefore, the idea of ML is not complicated; rather, it is an imitation of human learning procedures. Because it is not the result of programming, ML processing is not based on causal logic but summarizations.

• From Machine Learning to Deep Learning

ML is a natural consequence of AI development. From 1950s to the beginning of 1970s, studies of AI were still in a "reasoning period" and filled with presumptions. At that time, people thought machines could be considered intelligent if they are endowed logical reasoning skills.

Representational projects of this time were the "Logic Theorist" program developed by A. Newell and H. Simon as well as the following "General Problem Solving" program. These projects have all achieved exhilarating results. For instance, "General Problem Solving" proved 38 theorems in *Principia Mathematica* written by Alfred North Whitehead and Bertrand Russell in 1952, and then all its 52 theorems in 1963. Particularly, its reasoning for number 2.85 was even more tactful than Whitehead and Russell's. For this reason, Newell and Simon won the Turing Award of 1975 for their work.

However, as research in this field continued to evolve, people became aware that logical reasoning skill alone was not enough for achieving AI. Edward Albert Feigenbaum believed that machines must possess knowledge to be considered intelligent. This was the time when Machine Learning began to emerge.

As mentioned earlier, in 1952, Arthur Samuel developed the checkers program which overturned the argument raised by others on machines' ultimate inferiority to humans in programing and learning modes. Samuel coined the term "Machine Learning" and defined it as "research study that can empower computers without manifesting algorithms."

But due to the overall diminishing of investment in AI between the mid 60s to the late 70s, all areas—whether theoretical research or computer hardware design—encountered severe impediments in their progress. ML also fell into stagnation at this time. Although Winston's Structural Learning System as well as Hayes Roth's Inductive Learning System had relatively greater progress, the concepts were never put into practical use. Neural network learning machines also ground to a halt for not achieving the expected results because of theoretical defects.

After surviving the AI Winter, ML welcomed its spring in the 80s. In 1981, Paul Werbos specifically described a multi-layer cognitive machine model in a neural network backpropagation (BP). This was also the starting point when BP, since it was introduced as "reverse mode of automatic differentiation" in 1970, began to make a real difference in the study of neural networking.

In 1985 and 1986, as new ideas actively arose, the speed in neural network development quickened once again. The concept of multi-parameter linear planning (MLP) using BP was raised, which later became the cornerstone for Deep Learning.

At the same time, in 1986, J. Ross Quinlan came up with the ML algorithm "Decision Tree," also known as ID3. This led to many following explorations in the field including ID4, the Regression Tree, and the CART algorithm, which are still popular in ML studies today.

The invention of the Support Vector Machine (SVM) was another major breakthrough. It enjoyed a dominant status in the field supported by rich theoretical and empirical evidence, and since then the ML study was divided by two schools, the Neural Network (NN) and the SVM. After the development of SVM with kernel functions at around 2000, the advantage of SVM became more prominent and even exceeded NN in many tasks that it used to dominate. Moreover, in comparison to NN, SVM can also benefit from its profound knowledge in convex optimization, margin generalization theory, and kernel functions, which will greatly help to enhance theoretical and practical improvement from multiple disciplines.

In 2006, Geoffrey Hinton, leading figure in neural network studies, popularized the Deep Learning algorithm, which extensively improved the capacity of a neural network and challenged SVM. The paper that G. Hinton and Ruslan Salakhutdinov published in top academic journal *Science* officially set off the craze of Deep Learning in academia and in industry. In 2015, Yann LeCun, Yoshua Bengio, and G. Hinton published a joint synthesis on Deep Learning in memory of the 60th anniversary of the AI concept's creation.

Deep Learning enables and improves the learning of computing models with multiple processing layers in indicating multi-layered abstract data from many aspects. The creation of Deep Learning was a real advancement in cognitive issues relating to imagery and voices as well as in empirical application, which has been brought to an unprecedentedly viable state, propelling AI to a whole new level.

- **Development and Application**

In the past 20 years, our abilities in collecting, storing, transmitting, and processing data have improved enormously, which resulted in massive cumulation of data in every aspect of human civilization desperate to be processed by an efficient computer algorithm. ML thus receives broad attention and ample opportunities under this context for perfectly accommodating this urgent requirement.

Today, we can find ML in multiple areas of studies in Computer Science, such as Multimedia, Computer Graphics (CG), Network Communication, Software Engineering, Computer Architecture, and IC design. In particular, ML has become one of the vital forces in applied technologies like Computer Vision and Natural Language Processing (NLP).

Additionally, ML offers significant technical support for interdisciplinary studies. For example, research in bioinformatics, which covers the entire procedure of observing life phenomena and discovering patterns, is trying to apply information technology to the process consisting of data collection, data management, data analysis, and others. Among these phases, data analysis has offered ML an arena in which to shine.

It is fair to say that ML is the locus of the movement from Statistical Analysis to Big Data and the device to excavate the "new petroleum." It can help to enhance data analysis efficiency and forecast and inspection accuracy, both in basic application domains (e.g., environmental monitoring, energy exploration, weather forecasting) and in business application domains (e.g., sales analysis, image analysis, storage management, cost control, and recommendation system).

In other words, ML has made the iteratively updating personalized recommendation simple and handy in everyday life.

Think about how search engines like Google and Baidu have utterly changed the people's lifestyles. Living in the time of the Internet, people are used to search for information about their destinations, look for suitable hotels and restaurants before travelling. This demonstrates the profound influence of ML on modern societies. As we know, Internet searches are based on data analysis aiming for required information from users. During this process, users' inquiry (input) and the result provided by the Internet (output) lie on the two ends of the track, while the connection in between is built depending on ML technology. We can even argue that ML has been the reason for Internet search's glory today.

The more complex the subject of Internet searching is, the more prominent the impact of ML technology has become. Both Google and Baidu are using the newest ML technology with image search. Companies including Google, Baidu, Facebook, and Yahoo are forming special research groups or even institutes named after ML. All these examples reveal that the development and application of ML technology are shaping the future of the Internet industry.

Finally, aside from being the source of innovation for intelligent data analysis, ML is also significant in facilitating people's understanding of "how does human learning take place" through building learning-related computing models. In mid 1980s, Pentti Kanerva didn't have the structure of the human brain in mind when he came up with the Spare Distributed Memory (SDM) model. But neuroscientists later found out that the Sparse Coding system of SDM is widespread in the visual, auditory, and piriform cortex. This was an inspiring discovery for studying some of our brain functions.

The motivation for the pursuit in natural science derives from people's curiosity about the origin of the universe, the essence of the myriad of things, the nature of life, and consciousness of the self. Obviously, the question "how does human learning take place" falls in the last category. In this light, ML is not only the central focus in the realm of information science but is also of interest to those in the realm of natural science.

2.2 Computer Vision

• The "Eyes" of AI

A crucial division in AI technology, Computer Vision (CV) is viewed as the "eyes" of the intelligent world. It endows computers the abilities to "see" and to "recognize" mimicking the visual system of humans, serving as the cognitive foundation for computers. More specifically, CV technology applies a camera in place of human eyes to enable segmenting, categorizing, identifying, tracking, judging capabilities in computers. It is an integrated AI system that can acquire information from data of two-dimensional and three-dimensional images.

Similar to the way visual organs, visual cortex, and other parts of the brain cooperate to accomplish the "seeing" process for humans, CV uses an imaging system as means of input and visual control system as means of processing and interpretation to let computers accomplish a series of tasks automatically—detecting external visual information, initiating relevant interpretation and reaction, achieving more sophisticated decision-making and self-directed actions. As one of the most frontier areas of study in AI, CV technology is most keenly focused and widely distributed in the industry.

CV is an interdisciplinary scientific technology that overlaps areas such as Computer Science and Engineering, Neurobiology, Physics, Signal Processing, Cognitive Science, Applied Mathematics, and Statistics. Based on high-performance computing, CV can quickly acquire and process a huge amount of data, making it easier to design, handle, and control information integration. Research in CV focuses on various topics including Object Detection, Semantic Segmentation, Motion and Tracking, Visual Question and Answering, etc.

Another related terminology is Machine Vision, which is the application of CV in industrial scenarios. Its purpose is to replace traditional human labor, improve productivity, and reduce costs. While CV focuses on quality analysis such as categorizing objects, MV focuses on quantity analysis such as measuring or locating. Also, the application scenarios of CV are relatively sophisticated with various categories of objects in irregular shapes and implicit patterns, whereas the scenarios of MV are relatively simple and fixed, with fewer categories and more explicit patterns. But the latter has higher requirements for accuracy and processing speed.

• The History of CV Development

Abundant theories and methodologies have been raised during the 40 years of CV development. Overall, the history can be divided into three phases represented Computational Vision, Multiple View Geometry Stratified 3D Reconstruction, and Learning Based Vision respectively.

The publication of David Marr's research on Computational Vision in the book *Vision* in 1982 marked the establishment of CV as an independent subject.

Marr's Computational Vision theory includes two arguments. First, the main function of human vision is to recover the visible surfaces in three-dimensional scenarios, also called 3D reconstruction. Second, the reconstructing process from 2D images to 3D structures can be achieved via computation, and Marr provided a complete theory and methodology for this computation. For this reason, Marr's Computational Vision is also referred to as 3D Reconstruction theory by some scholars.

The theory maintains that the process of reconstructing an object's 3D structure from its 2D images consists of three stages. The first is the theoretical stage, which means what type of principle is needed for the process. The proper principle, according to Marr, is the principle formed during the imaging process by the inherent nature of the scenario. The second is the presentational and algorithm stage, which means how to carry out the computation. The third is the application stage.

Marr illustrated the presentational and algorithm stage in detail. He argued that the 2D description of an object, or the sketch, needs to be first reconstructed to a 2.5D description before to a 3D description. A sketch is characterized by the primitive features such as zero-crossing as in Laplacian of Gaussian (LoG), short line segment, and endpoint. A 2.5D description is a crude portrayal of the object, like its normal vector in the observer's coordinate system. A 3D description is a portrayal of the object in its own coordinate system, like the portrayal of a sphere using its center as origin.

The influence of Marr's Computational Vision is profound and pervasive. His Layered 3D Reconstruction framework is still the mainstream approach in the field of CV.

Starting from the 80s, CV attracted worldwide attention. New theories were actively raised in alternation thanks to two factors. On the one hand, the application focus of CV switched from industrial domain with high demands for precision and robustness to areas that only required visual effects, such as teleconferencing, archeology, Virtual Reality (VR), cameras and surveillance. On the other hand, scientists realized that the precision and robustness of 3D reconstruction can be sufficiently improved with Layered 3D Reconstruction in Multiple View Geometry. During this time, OCR and smart cameras came out and further intrigued the dissemination and application of CV-related technologies.

Halfway through the decade, CV has progressed rapidly. Original concepts, theories, and methodologies of the Active Vision framework, cognitive-feature-group-object-recognition-theory-framework, and others were raised consecutively.

In the following ten years, CV was popularized in industrial environments. Studies that benefited the most at this time were the Multiple View Geometry-based vision theories. At the beginning of the 90s, image technology corporations were established and developed the first-generation image processing products. Since then, CV-related technologies became closely connected with industrial production, which in turn sponsored the expansion of CV territory. Hundreds of companies started to sell CV systems, and so a complete CV industry was formed. With the simultaneous booming of sensors and control structures, the producing costs in the CV industry continued to drop.

Into the 21st century, we see a tendency of intensifying mutual influences between CV and CG, and Image-based Rendering (IBR) is becoming a new research hotspot. Studies of algorithms in high-efficient global optimum solutions for complex issues are in progress. Faster 3D visual scanning and thermographic systems are created. CV software and hardware products are penetrating and expanding in every stage in the production process. Serving both in the fundamental AI industries and in the upstream industries like electronics and automobiles, the prospect of CV is highly positive with vast space for further development.

• Extensive Use of CV

CV is an essential component in new infrastructure creation. It occupied 34.9% of the AI market in 2018, ranking first, and it also displayed an overwhelming advantage in investment and financing. Supported by its maturing technology, the CV market in China is growing at a remarkable speed. According to iiMedia Research data, the overall market size of the Chinese CV market in 2018 has reached 15.5 billion RMB, outnumbering that in 2017 by 8.7 billion RMB, with a Compound Annual Growth Rate (CAGR) of over 100%.

The development of CV industry is dually driven by two factors—the market and technology.

The market-driven perspective refers to the changing of the market driving force since the demographic dividend disappeared as machinery replaced human labor and brought enormous economic benefit to the society. Take the industrial vision system for example. In developed countries, a regular US$10,000 industrial vision system can do the same work as three workers with an annual salary of US$20,000. But it's payback period is much shorter, and the subsequent maintenance fee is much lower. The economic advantage of machinery over human labor is thus apparent.

We are living in a time of videos exploding with countless data waiting to be processed. Around 70% of our cerebral cortex is processing the content of what we see, or the visual information. Before the CV's advent, images have been a "black-box" for machines just like how they would have been for humans if their visual capability were deprived. CV enabled computers to "see," "understand," and then process visuals.

Videos are delivered to us in all possible forms via almost all available apps, with the standard of 5G succeeding 4G. This accessibility and speed resulted in an exponential growth in video data. Therefore, further advancement in AC technology is inevitable if we want more precise management of this new data category.

The technology-driven perspective refers to the CV industry development stimulated by new generation information communication technologies like 5G and AI technologies like Deep Learning.

On the one hand, basic CV practices such as facial recognition and optical character recognition (OCR) already existed in the 4G era. Now, with the universalization of 5G, higher demands for the CV products to be high-speed, wireless, and movable can be met as well. On the other hand, a new algorithm with more powerful computing capacity combined with big data of the field will allow AI technology to fit in a broader range of scenarios. It can increase the efficiency and reduce the costs in areas such as security, industrial production, and medical imagery diagnosis.

This will necessarily promote the application of CV. The current application focus in China prioritizes security, finance, and the Internet, while the focus in other countries prioritizes consumption, visual robots, and smart driving.

The cause for this distinction is threefold. First, the difference in market need. Security and digital finance are the most pivotal scenarios in China's leading relevant industries' growth. Second, the difference in developmental period and phase. In western countries, the study and practice of CV technology started several decades ago and have now entered a relatively sustainable developmental stage. In contrast, CV companies in China were only founded and developed since 2010—which is also fortunate for China given the global environment of flourishing large-scale visual technology application and Internet explosion. Third, the difference in market emphasis. The preference of IC and hardware over software algorithms in Western markets leads to its emphasis on IC development and market monopoly, whereas the Chinese market focuses mainly on transforming technical knowledge and engineering experiences into vertical solutions that address the issues in business.

The most typical application in CV is undoubtedly facial recognition systems, which are now widely used in various scenarios such as retailing, finance, and livelihood with increasing precision based on Deep Learning technology.

Deep Learning is adept in acquiring the robust changing patterns of human facial features based on massive training data. It does not require programing on specific features regarding different types of intrapersonal variations (e.g., light, pose, expression, age, etc.), for it can acquire them directly from training data. The Convolutional Neural Network (CNN) is the most commonly used Deep Learning approach in facial recognition because it is highly stable with different kinds of deformations like translation, scaling, tilting, and others. In addition, it can still achieve high identification rates with large image samples.

The process of facial recognition involves the steps of facial detection, facial alignment, and facial recognition. The specific procedure is as follows.

The algorithm first detects a human facial area within an image. It then aligns with the key feature points on the face it has detected, meaning it will make sure the facial features such as eyes and mouth have the same coordinate points in the image. After that, the neural network will draw upon the facial features for consequential training and build up a deployment model. Collect the features drawn from each facial model to get an eigenvector. Calculate the cosine distance of the eigenvector using cosine similarity. Those models with cosine distances smaller than the threshold value are deemed as the same person.

Aside from facial recognition, object digitalization via OCR is another representational application in CV. OCR is a technology in recognizing texts within images that can be applied in license plate recognition (LPR), ID card recognition, and passport recognition.

One representative in object digitalization via OCR is VisionSeed under the Chinese tech company Tencent, a Chinese multinational entertainment conglomerate and holdings company and the largest gaming company in the world. Based on its theoretical expertise in OCR and rich empirical experiences, the corporation self-developed a versatile OCR engine of high precision with two core algorithms for recognizing multi-dimensional texts in random forms and for recognizing semantic-comprehension-integrated texts. It conquered many classical obstructions in the field regarding text deformation, intensive configuration, complex background interference, handwriting, small text, and unclear text using the millions of training sets constructed by self-developed emulated data algorithms.

In order to fully validate the capacity of the algorithm, VisionSeed OCR has tested it with several thousands of images that encompass dozens of scenarios (e.g., files, road signs, books, test papers, courier sheets) with an accuracy rate of 95%. The company continued to develop OCR technologies for over 50 vertical categories (e.g., documentation, education, banknotes) with an accuracy rate of 98% on key segments. They are then applied in finance, insurance, accounting, logistic, education, and other fields through Tencent Cloud OCR, improving info-entry speed by 90% and saving human-labor-entry cost to a significant extent.

2.3 Natural Language Processing

- **Language Processing Machine**

After Turing invented the Turing Test that provoked thought in natural language processing in 1950s, and after a series of upgrades in fundamental technology system succeeding Expert System, Statistic Machine Learning, and Deep Learning for around half a century. Natural language processing skill has matured considerably today. It is now playing an essential role both in academic research and in practical applications.

Natural language refers to languages that people use every day like Mandarin, English, and French. These are not artificial languages, but languages developed from the evolution of civilization. They are important tools for learning and living in human societies. In human history, information recorded and inherited both in oral and written forms of natural languages has occupied above 80% of all information. In terms of computer application, almost 85% of natural language is used for language information processing, while the remaining 15% is used for math calculation (around 10%) and process control (less than 5%).

Natural Language Processing (NLP) transfers languages people use for daily communication into machine languages that computers can understand. It is an algorithm framework that studies language ability, and so the area of NLP is interdisciplinary, encompassing both linguistic and CS studies for the purpose of attaining human-computer information exchange. Development in NLP appeals simultaneously to the AI, CS, and linguistic academic circles.

The specific manifestation of NLP mainly includes machine interpretation, text summarization, proofreading, information extraction, speech synthesis, and speech recognition. NLP consists of two procedures, natural language comprehension and natural language production. The former indicates comprehending the content of natural language texts, and the latter indicates conveying specified implications with natural language texts.

Basically, NLP procedure follows five steps.

First, corpus obtainment. Second, process pretreatment, which includes several minor steps such as corpus cleaning, word separating, part-of-speech tagging, and stop-word removing. Third, characterization, or vectorization, which means representing characters and words in measurable forms (vectors) to better indicate the connections between different words. Fourth, model training, which includes supervised, semi-supervised, and unsupervised forms that can be selected, based on application requirements. Fifth, effect assessment. The usual assessment indicators are Precision, Recall, and F-Measure. Precision is the measurement standard for the retrieval system's accuracy rate. Recall is the measurement standard for the system's recall rate. F-Measure is the index that reflects on the overall accuracy rate and recall rate. The test method is deemed effective when the F-Measure is high.

Bill Gates once said, "Language understanding is the crown jewel of AI." In other words, whoever dominates NLP skills of a superior level dominates the future race in AI.

- **The Path to Prosperity**

The journey of NLP was not all smooth sailing.

During the first twenty years following the introduction of the Turing Test, studies in the field were mainly guided by a Rationalist rule-based approach and the belief that NLP parallels human language-learning-process.

However, there are some unavoidable limitations with the rule-based approach. The rules cannot cover all sentences, and it also requires the developers to have expertise in CS and linguistics. For these reasons, many sophisticated questions about NLP were still left unsolved at this point and application was unattainable.

After the 70s, with the support of the fast-growing Internet, larger corpus, more advanced hardware, and the transition of the guiding ideology from Rationalism to Empiricism, a statistics-based approach gradually replaced the rule-based approach in NLP studies. The generating force behind this transition was Frederick Jelinek and his team in the IBM Watson Scientific Computing Laboratory. With a statistics-based approach, they enhanced speech recognition precision from 70% to 90%. This substantial breakthrough in math-model-and statistics-based approaches made practical application possible.

NLP has flourished since the 1990s. On the MT Summit IV held in Kobe, Japan in July 1993, Dr. William John Hutchins pointed out in his report that machine interpretation has entered a new era since 1989. It featured the introduction of the corpus approach, which included a statistics approach, an empirical-based approach, and the approach of the language knowledge base transformed from the corpus via the corpus processing method. It was a revolution in machine interpretation studies, based on large-scale authentic texts processing, which is destined to bring NLP to a brand-new level.

Especially in the last five years of the 1990s and the beginning of the 21st century, changes in NLP occurred on an unprecedented scale and this was manifested mainly in the following three aspects.

First, the probability and data-driven approach became the standard approach in NLP. It was applied in syntactic analysis, part-of-speech tagging, reference resolution, and discourse processing algorithms. The assessment method was borrowed from the disciplines of speech recognition and information retrieval.

Second, commercial development in NLP sub-disciplines such as speech recognition, spellcheck, and grammar check became viable due to computer's enhanced speed and storage capacity. Algorithms in speech and language processing were also applied in Augmentative and Alternative Communication (AAC).

Third, the development of grid technology that provided tremendous impetus to NLP. For example, the development of the World Wide Web (WWW), constructed by natural languages, highlights the demand in online information retrieval and extraction and is stimulating the advancement of data mining technology along the way. It can be predicted that research in NLP will play an increasingly important role in guiding the development of WWW in the future.

Encouraged by the achievement in image recognition and speech recognition, scientists are now bringing Deep Learning into NLP studies. In 2013, the attempt reached a climax with Word2vec's success in areas like Machine Interpretation, Question Answering System (QA System), and Reading Comprehension. As a multi-layered neural network, Deep Learning designs and trains its neural networks based on data that has experienced non-linear change layer after layer between the two ends of input and output. At present, Recurrent Neural Network (RNN) has become the most popular approach in NLP, while models like Gated Recurrent Unit (GRU) and Long Short-Term-Memory (LSTM) keep prompting new interest and topics in NLP.

2.4 Expert System and Knowledge Engineering

- **From Expert System to Knowledge Engineering**

The world's first Expert System DENDRAL was invented in 1965. Within 30 years, the technology and its application have proceeded rapidly. Especially since the mid-80s, different kinds of Expert System products based on Knowledge Engineering have added new fuel to AI growth.

An expert is someone with certain skills or professional knowledge in academia or technical areas. But other than theoretical and empirical expertise, experts should also have unique ways of thinking that allow them to analyze and solve problems.

With this understanding, it can be inferred that the Expert System (also called Expert Counseling System) is a computer (software) system that deals with sophisticated practical questions like human experts.

As a specific knowledge-based system, Expert System building requires the following steps: Knowledge Acquisition, which means collecting, organizing, and synthesizing expert-level knowledge from human experts or actual problems; Knowledge Representation, which means conveying acquired knowledge in certain forms and storing it in computers; Knowledge Base, which means knowledge organization and management; Knowledge Construction, Maintenance, and Application, which means a series of techniques and methods in knowledge inferencing.

In general, Knowledge Engineering is an area of study on knowledge processing approaches that derives from and supports the Expert System.

• Building an Expert System

In concept, Expert Systems (ES) are all grounded in the same structural model that includes a Knowledge Base, an Inference Engine, a Dynamic Database, Human Machine Interaction (HMI), an Interpretation Module, and a Knowledge Base Management System. Among them, the Knowledge Base and the Inference Engine are the most fundamental.

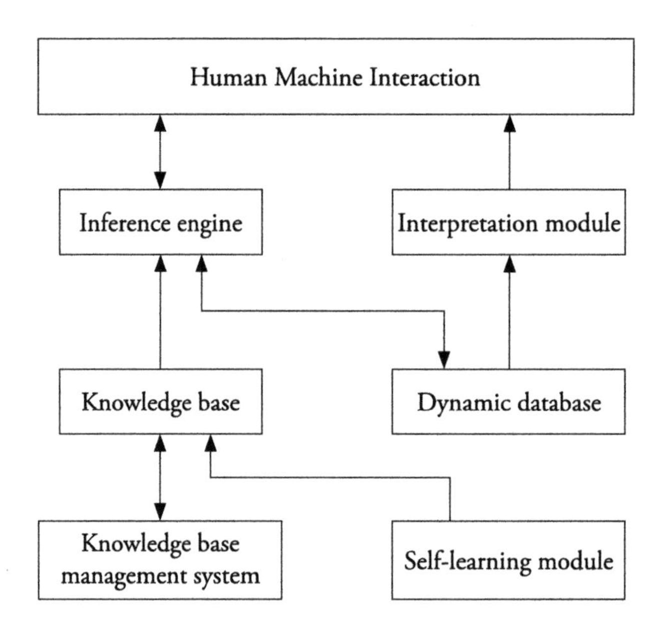

Knowledge Base is a certain form of knowledge set stored in computers. The most often applied form is multiple separate files stored in an external medium, which will be transferred to memory when the ES operates. The content of the Knowledge Base can be generally divided into expert knowledge, domain knowledge, and meta-knowledge—knowledge about controlling and managing knowledge. It is structured according to the representation, nature, level, and content of information.

Knowledge in an ES is most represented by the production rule (other representations of AI knowledge also include framework, semantic network, etc.). A production rule comes in the form of "IF…THEN…" statement as the conditionals in BASIC and other coding systems. "IF" is followed by the condition (antecedent or *P*), and "THEN" is followed by the result (consequence or *Q*). Both condition and result can be combined through logic operations AND, OR, and NOT. The production rule here is simple, IF the condition is satisfied, THEN the consequential statement is executed.

The Inference Engine is a program that enables (machine) reasoning. It repeatedly applies the rules in the Knowledge Base to given conditions and existing information of an issue to gain new conclusions and solutions. There are two inferencing modes that Inference Engine works in: forward chaining and backward chaining.

Forward chaining starts with looking for rules with premises that correspond with facts or statements in the Knowledge Base. It will then apply conflict-elimination strategy to select and execute one of the available rules which will then change the content in the original Knowledge Base. The engine will iterate until the facts in Knowledge Base correspond with the target or when no corresponding rules are found.

Backward chaining starts with looking for rules with consequences that can reach the selected target. If the premise of the rule corresponds with the facts in Knowledge Base, the engine will conclude. Otherwise, it will continue to look for rules that can reach the new sub-target, which is the premise of the previous rule. The engine will iterate until the premise of the rule corresponds with the facts in the Knowledge Base or when no corresponding rules are found. In that case, the program will request a conversation with the user for necessary facts.

The Inference Engine solves problems by deduction based on information in the Knowledge Base. We can think of it as the realization of expert thinking, analyzing, and disentangling in computers.

Together, Knowledge Base and Inference Engine build up the framework of Expert System. Of course, different knowledge representation has different inferencing modes, and so the Inference Engine's iteration and proficiency depend on the knowledge, the Knowledge Base structure, and the algorithm the engine selects.

Dynamic Database (also called global database, integrated database, working memory, and blackboard) is where original facts, inferencing results, and control information are stored—or we can say it is a set of these data. The term "dynamic" derives from the fact that a Dynamic Database only generates, changes, and revokes during program execution. However, although it is also termed "database," a Dynamic Database is not a database in the general sense.

The Human Machine Interaction (HMI) indicates the user interface, the ultimate point of human-ES communication. On the one hand, users can raise or answer questions and provide original data and facts for the system. On the other hand, system can raise or answer questions, show results, and provide explanations for the iteration and result for the users.

The Interpretation Module oversees explaining iterations and results for users. It answers the "why" questions about system behavior from users during the deduction, and it answers the "how" questions about the deduction process after it concludes.

The Knowledge Base Management System is the supporting software of the Knowledge Base. Its function resembles that of sub-database management system within a database, including database construction, deletion, and restructuring; knowledge acquisition (input and editing), maintenance, inquiry, update, and inspection (consistency, redundancy, and completion).

The Knowledge Base Management System is mainly used in the developmental phase of an Expert System, but it is also often used during the operating process to manage the Knowledge Base. Therefore, the lifespan of Knowledge Base Management System is the same as that of the corresponding Expert System. Generally, users of the Knowledge Base Management System are developers of the system including experts in the area and computer scientists (knowledge engineers), while users of the finished product of Expert System are specialists in different areas.

- **Development and Application**

The world's first ES, DENDRAL, was invented by Dr. Edward Albert Feigenbaum and a team of research fellows in 1965 at Stanford University. It was an ES for chemistry that can deduce molecular structures of chemical compounds based on their molecular formulas and mass spectrometry data. The success of DENDRAL gave strong impetus to AI scientists and new hope to the field, as it marked the transition of AI from lab research to empirical use. Now that ES has built up its own research territory, it changed the concentration of AI studies from deduction to knowledge.

Up to the 1970s, ES became more widely acknowledged. A few outstanding ESs, especially in medication, came out around halfway through the decade. One of the most typical among them was MYCIN, invented by Edward H. Shortliffe and his team for the purpose of diagnosing and treating infectious diseases. They started in 1972 and finished the basic framework in 1974. Afterwards, they continued to improve and expand the program and turned it into the first comprehensive ES. Aside from the original functions it was designed with, MYCIN was easy to operate, be comprehended, revised, and expanded. For example, it could engage in conversations with users and answer questions through natural languages as well as learn new medical knowledge guided by human experts. It was the first to apply the Knowledge Base and likelihood reasoning technology, and it served as the foundation for many later ESs.

In 1977, at the Fifth International Joint Conference on Artificial Intelligence (IJCAI), Dr. Edward Albert Feigenbaum, creator of ES, illustrated this idea in detail and introduced the concept of Knowledge Engineering in his paper "The Art of Artificial Intelligence: themes and case studies of knowledge engineering."

So far, ES theories were basically accomplished, and Knowledge Engineering, a new study of ES theory, methodology, and technology was born.

ESs in the first half of the 70s mainly addressed the issues of data analysis (e.g., DENDRAL, PROSPECTOR, HEARSAY, etc.) and fault diagnosis (e.g., MYCIN, CASNET, INTERNIST, etc.) The problems they dealt with were mostly decomposable.

ESs in the latter half of the 70s were more diverse in category. These included:

- KBVLSI—Very Large Scale Integration Circuits (VLSI)
- PSI—automatic programming and other designing systems
- MOLGEN—genetic experiment systems
- NOAH—robot movement planning and other planning systems
- GUIDOON—infectious disease diagnosing and curing systems
- STEAMER—steam power operation teaching and other education systems
- IW—military conflict prediction and other forecasting systems

At the same time, programming languages and instruments suitable for ES development were produced, which helped to pave the way for the popularization of ES in industrial production in the 1980s.

Up to the 1980s, industrial applications of ESs in the West included:

- CAD/CAM and engineering design
- Machine defective diagnosis and maintenance
- Manufacturing process control
- Dispatching and manufacturing management
- Energy management
- Quality insurance
- Petroleum and resource exploration
- Electricity and nuclear facilities
- Welding

2.5 Robots

• What is a Robot?

Broadly speaking, robots are all mechanisms that imitate human behavior or thinking or that imitate other organisms (e.g., robot dogs, robot cats, etc.). But there are more various and detailed nuanced definitions of the term in academia and professions. Even computer programs can be called robots too (e.g., Web crawler)

The International Organization for Standardization (ISO) applies the definition of a robot as given by the Robot Institute of America (RIA): a robot is a programmable versatile operator; or a specialized system that can be changed by computers and have programable functions for executing different tasks. It is usually comprised of an execution structure, a driving device, an inspecting device, a control system, and complex machinery. The most important keywords in this definition are "programmable," "versatile," and "operator."

"Programmable" indicates that the program of a robot is not one-time coding but can be designed any number of times according to need. Many electronic devices we use in everyday life are designed with programmable computer chips. For example, you can enter a program in the chip of a digital alarm, which orders it to play *Auld Lang Syne* when the alarm goes off. However, a program like this cannot be changed, and it doesn't allow new programs to be entered either. In contrast, the programs of robots can be re-entered, changed, added, and deleted based on the users' needs. A robot can also have multiple programs and conduct different tasks in random order. Of course, in order to re-program, a robot must have a computer that allows new orders and information to be entered. This computer can be a "carry-on," meaning the control board is installed on the robot. But it can also be remote, meaning that it can be anywhere away from the robot as long as information exchange between the two remains unimpeded.

"Versatile" indicates that the robots are multi-functional and can be used in various situations. For example, robots for laser cutting can also be used for welding, spray painting, or arranging devices with just a slight change of its terminal tool.

"Operator" indicates that when a robot is at work, it needs a system that moves the working subject. While the difference between a robot and other automatic machines lies in a programmable and versatile coding, the difference between a robot and a computer lies in the operating system.

In short, a robot is a sophisticated intelligent machine that integrates mechanics, electricity, computers, sensors, control technology, AI, bionics, and other disciplines.

• The History

In ancient Greece and Rome, primitive robots existed in various interesting forms. Throw eight coins into the mouth of the stone griffin, and holy water will come out from its eyes. When Pythia light up the holy flame before a temple, its gates will open by itself—or, as modern engineers will term it, "automatically."

The first "reliable" record of the early robots is found in *The Iliad*, which tells the story of a woman made of gold helping the god of blacksmith Hephaestus. These heavily romanticized tales about the ancestors of robots convey a strong sense of curiosity for the imagined world with artifacts that can behave like a human.

Robots became part of human reality along with the need of automatic production and scientific research. The robot Elektro built by the Westinghouse Electric Corporation was exhibited at the 1939 New York World's Fair for the first time. It could speak simple sentences, walk, and smoke. But it couldn't help people with chores.

Modern robots originated between the 1940s and 1950s when many national laboratories in America began to experiment in this field. During WWII, an easy remote control was developed for radioactive material production and processing by replacing human operators with mechanic handles.

After this, Oak Ridge National Laboratory (ORNL) and Argonne National Laboratory (ANL) started to develop remote handles for carrying radioactive materials. In 1948, Master/slave remote handle was born and set a precedent for modern robot manufacturing. Later, MIT Radiation Laboratory successfully developed the Computer Numerical Control (CNC) in 1953 which combined the technologies of servomechanism and the latest digital computer. The Orthogonal Cutting model will be entered into the machine through punched tapes in the form of numbers, and the servo axis will enact the cutting based on the model trajectory.

The 1950s was the era when robots turned practical. In 1954, George Charles Devol designed and built the first robot experimental device, published the article *Universal Industrial Robot Suitable for Repeated Operation* and obtained a patent for it. This design has artfully connected the articulated mechanical linkage of the remote control to the servo axis of the CNC. The mechanical handle can enact the pre-scheduled activities independently after the program is entered. The robot can also receive teach-ins and carry out various simple tasks. During the teach-in, the operator will guide the mechanical handle through the locations needed for the task in order, which will be recorded in the digital memory. The different articulations of the robot will then reproduce the location sequence with servomechanism. "programmable" and "teach-in reproduction" are thus the two key feature functions for this type of robot.

In 1960, Consolidated Control Corp. developed the first sample robot based on Devol's invention and founded Unimation which produced industrial robots under the brand name Unimate. At the same time, American Machine and Foundry (AMF) produced another kind of programmable robot, Versatran. With "teach-in reproduction," the two types of robots successfully replaced human labor in transporting, welding, and paint spraying on the vehicle assembly line, displaying phenomenal economic benefits, reliability, and flexibility that appealed to other countries. In Japan, for example, the robot industry has also achieved extraordinary progress since 1968.

The Experimental walking vehicle that GE developed for US Army in 1969 was an important milestone in the history of robots. The enormously complicated control system was beyond human capacity, and so it propelled the progression in auto-control studies. The main theme was the four-legged gear's high demand for flexibility. Boston Arm was born in the same year, and Stanford Arm in the next year with a camera and computer control unit. These are pivot points in the developmental progress for robot studies where machinery was used for robot control systems.

In 1970, the first national robot conference was held in the US. In 1971, an industrial robot association was established in Japan for promoting robot application. The first computer-control robot that came out afterwards was a T3, which could lift the weight above 100 pounds and track the workpieces on the assembly line. It was praised as "future tool."

Following this, more robots were designed with different coordinate systems and structures to accommodate various scenarios. But it was not until the 80s when Large Scale Integration (LSI) technology and microcomputers became popular and efficiently improved the robot controllability and reduced its cost that hundred types of robots were put into use.

Robots by this time already possessed basic capabilities in perceiving and responding. More countries with developed manufacturing industries began to apply robots in vehicle assembly lines and other industrial productions.

The next stop for robots was to become smart. The new robots were designed with multiple sensors that could integrate information received through various sources. They worked very well in adapting to the changing environment, learning, and auto-controlling. The key technologies involved in smart robot development, succeeding programmable teach-in, and reproduction robots, are multi-sensor information integration, navigation and positioning, routine planning, vision control, and HCI.

In the 21st century, with the continuous improvement of labor costs and technology, countries have carried out the transformation and upgrading of the manufacturing industry, and there is an upsurge of robots replacing humans. At the same time, with the rapid development of AI, service robots are beginning to enter the life of ordinary families. Many robot technology companies in the world are vigorously developing robot technology, and robots are getting ever closer to organic life.

CHAPTER 3

AI Applications Becoming Ubiquitous

3.1 AI in Medication

- **What Can AI Pharmacy Do?**

It is now a world-acknowledged fact that AI is widely applied in the healthcare sector. Aside from conducting diagnosis and operations, managing treatment resources, and offering medical insurance, AI is also showing an advantage in innovating disease and drug research and is becoming its main driving force in recent years.

Generally, the development of a drug undergoes two phases: drug discovery and clinical trials.

Drug discovery follows the procedure of building an assumption of a disease, finding the drug target, designing the compound, and moving on to preclinical research. The traditional drug developing process involves a huge number of simulation tests that has a high demand on time and cost but little promise of a success rate. According to the data provided by *Nature* in 2014, the average cost for developing a new drug is US$2.6 billion, the average time taken is 10 years, but with a success rate below 10%.

Every step (i.e., drug target discovery, compound design, hit selection, lead compound improvement, candidate compound confirmation) during the process faces a high risk of elimination. With drug target discovery, for example, effective chemical molecules need to be found among several hundreds of molecules through repeated screening. In addition, the convergence of human minds may result in the production of similar drugs that aim at one drug target, and this may cause unnecessary patent litigation. Finally, although thousands of compounds may be needed for screening, only very few of them will pass it and be selected for development into a new drug.

However, AI technology can reduce the costs by exploring the underlying connection between disease, genes, and drugs. Based on data of disorder metabolism, large-scale genome recognition, proteomics, and metabolomics, it can conduct high-flux virtual screening with candidate compounds and search for the connections between drugs, disorders, and genes. This way, it can improve the efficiency and success rate of drug development.

To be more specific, AI helps scientists to search for literature, patents, and clinical results that come in huge numbers through text analysis. It will reveal potential connections regarding signaling pathways, proteins, and mechanisms that are relevant to the disease. Following this, it will further propose new testable assumptions for new mechanisms and new drug targets. For example, IBM Watson has applied AI technology in detecting the connection between several thousand genes with ALS, a rare disease caused by specific genes. They found five genes related to ALS and successfully propelled our understanding of ALS (there were three more genes that had been already found in previous research).

By conducting virtual screenings with candidate compounds, AI helps scientists identify highly active compounds that can promise a higher success rate and save time in drug screening. For example, Atomwise has applied Convolutional Neural Network (CNN) AtomNet in support of complementary medicine development for structure-based drugs. By having AI analyze the process of simulated development of a drug database, it can anticipate potential candidate drugs, assess the developmental risks for the new drug, and predict the efficiency of the final product. Astellas Pharma and NuMedii also seek biomarkers for selecting new candidate drugs and anticipating diseases with neural-network-based algorithms in their collaborative research.

Clinical trials are the most time consuming and costly period in the drug development process. Clinical trials commonly consist of multiple phases, including phase I (screening for safety), phase II (establishing preliminary efficacy), and phase III (confirming safety and efficacy on a large scale).

In traditional clinical trials, the cost for recruiting patients is high and information symmetry is difficult to guarantee. According to a survey by CB Insights, the biggest reason for clinical trial postponement (up to 80% of all trials) is the failure in finding ideal clinical volunteers in time.

Another major issue that should be taken into consideration is the clinical agreement. If volunteers failed to strictly abide by the agreement, their relevant data would need to be deleted from the set; otherwise, the false data can cause a serious distortion to the results of the trial. What's of equal importance assuring the result's accuracy is making sure the volunteers take the right drugs at the proper times.

These challenges are made easy with AI technology. Effective information from the patients' medical records will be automatically extracted and matched with ongoing clinical trials. This way, the recruiting process will no longer be cumbersome.

When it comes to the difficulty in keeping track of medication compliance in patients, AI technology can achieve continuous surveillance by tracking the intake of drugs and compliance with sensors and facial recognition devices. The open-source framework ResearchKit and CareKit created by Apple, for example, can not only help recruit volunteers before the trial but also remotely monitor patients' health conditions and daily lives during the process.

Given the benefits and potential of AI technology in pharmacy, and the fact that people are getting more and more used to the achievements in the field of AI studies (i.e., AI algorithm discovering a new powerful antibiotic), the question is, why hasn't AI pharmaceutical industry yet to be fully exploited?

It is true that the future of AI pharmacy appears highly promising. During the COVID-19 pandemic, AI technology has done a remarkable job in assisting case diagnosis and management. With the big data platform co-established by Alibaba Cloud and the Global Health Drug Discovery Institute (GHDDI) in China and AlphaFold by DeepMind in Europe and America, data collection regarding previous drugs for coronaviruses and anticipation of possible structures for the current one has progressed steadily.

Despite the difference brought by advancing technologies, the costs of new drug development are still increasing partly due to publicity stunts and competition derived from increasing investment and human resources. Therefore, the positive prospect of AI pharmacy has never been realized.

The conclusion we can draw from the status quo of AI pharmacy is that it is still going to remain mildly influential due to the limitations of AI. Right now, its chief assignment is still learning to build models from a knowledge base. The more data there is in the knowledge base, the smarter the technology becomes.

Succeeding the first two generations of GPT models, which included 117 million and 15 billion parameters respectively, the GPT-3 model has multiplied this number by over 100 times and achieved an astounding 175 billion, more than 10 times the parameter number of NLP model, its largest counterpart of the same category. It realizes intelligence by analyzing 500 billion words and can compose a coherent dialogue of any topic with only a few words' hint. But the data size is also a confinement for the model's potential.

Generally, high-quality data regarding imitating specific behaviors needs to be entered into the model for it to achieve extraordinary functions. In games such as *go*, it means there must be a clear parameter for every step. But this is a lot harder to achieve in a real life environment that is less predictable. This issue is causing many difficulties for AI when applied in actual scenarios.

Another reason that AI failed to support governments of France, the US, the UK, and other countries during the pandemic in building an efficient contact-tracking system, which lacks the necessary "raw material," the data for tracking and tracing the sources of COVID-19 cases.

Only sufficient input from science can ensure better output in science. The experiences we've had with AI pharmacy so far indicate that although AI proves to be useful in developing urgent drugs and improving healthcare, support from other fields such as biology and chemistry is still needed, whether in terms of generating Reinforcement Learning (RL) or realizing what's promised in the future of quantum computation.

With the changing in demographic structures and the emerging of vaccines and antibiotics in modern societies, chronic diseases are replacing infectious diseases in causing the highest threat for people's health. However, the existing healthcare systems are still based on infectious and acute disease treating experiences. It is less fruitful in dealing with chronic diseases and is under increasing pressure from high medical costs.

What AI can help with the situation are: improve early inspection accuracy, reinforce risk control, reduce treatment costs, assist patient personal health management, and enhance treatment efficacy. It can achieve these targets by imitating human brain functioning, with integrated DL, data modeling, and large-scale GPUs.

To sum up, although there are still some uncertainties in developing AI pharmacy into a mature industry that can replace biological and chemical pharmacy, it doesn't mean the golden age for this new approach in drug production isn't coming. In the future, it will bring revolutionary changes to ways of identifying drug targets, designing new molecules, providing personalized treatment, and anticipating clinical trial results.

• AI and Mental Healthcare

Aside from the aspects that are accessible and familiar to all members in our society, AI also plays an important role in areas that are less popular but indispensable such as the diagnosis of mental disorders.

Right now, there are around 16 million patients in China suffering from severe mental illnesses. According to the WHO, by 2020, depression had already become the second largest threat to human health. Around 30% of the patients show no response to medication; while among the responsive population, only 30% can obtain clinical remission.

Mental disorder diagnosis needs to refer to the *International Classification of Diseases* (ICD), the *Diagnostic and Statistical Manual of Mental Disorders* (DSM), and a survey done by experienced doctors. It cannot be confirmed based on blood tests, CT scans of the brain, or biopsies alone. Therefore, there is no easy way of assessing the damage that coronavirus does to people's mental health even though psychoanalysts are worried about this not only being a hypothesis.

The fact that there is no reliable biomarker that can be used in diagnosing mental disorders has prevented and discouraged many studies in psychiatry. It has made the diagnosis of mental illnesses slow, difficult, subjective, and even trickier for scientists to comprehend the real essence of mental disorders or think of better treatment for them.

Nevertheless, there are cues that a psychiatrist can look for when detecting their patients. For example, language.

In 1908, Swiss psychiatrist Paul Eugen Bleuler introduced the term "schizophrenia" for a disease he and his colleagues had been working on. He noticed that the symptom of this disease is "revealed through language," but he also added that "disorder lies not in the language itself but in what it conveys."

Bleuler was among the first scholars to pay attention to the negative symptoms of schizophrenia, meaning symptoms not found in unaffected people. These symptoms are not so obvious as positive symptoms, which refer to accessory symptoms such as delusions. One of the most common negative symptoms is stuttering or other language disorders. Schizophrenia patients are inclined to speak very little with many obscure, repeated, and rigid phrases. This is what psychiatrists call low semantic density.

Low semantic density is a dangerous sign for schizophrenia. Some research show people in the high-risk group generally don't use possessive pronouns such as "mine," "his," or "ours." Semantic identification is thus a possible breaking point for schizophrenia diagnosis.

According to the *Digital 2021: Global Overview Report* released by DataReportal, as of January 2021, among the world's 7.83 billion people, there were 5.22 billion mobile phone users, 4.66 billion Internet users, and 4.2 billion social media users. It means the ubiquitous smartphones and social media make people's language easier to record, digitize, and analyze than ever before.

The truth is, the enormous quantity of data related to healthcare we acquire through mobile devices may be far more valuable than the definitions for disorders scientists agree upon through traditional means including physical examinations, lab tests, and radiology. In fact, more and more researchers are trying to find patterns in depression, anxiety disorder, bipolar disorder, and other mental illnesses from data produced by people, such as the language we choose, the ways we sleep, and the frequency we call our friends. This data and the analysis of it are called phenotype data.

The connection between individual and digital science in phenotype data influences all stages on the spectrum including diagnosis, treatment, and chronic disease management. Phenotype data in psychiatry can help better track the biomarkers (e.g., emotions, motions, heart rate, sleep) in patients' daily lives and relate them with clinical symptoms for more efficient tests.

The benefits of using AI-based phenotype data in comparison to relying on psychiatrists' expertise, experiences, and personal opinions are evident. With disease prediction, for example, the current most advanced application of phenotype data in use is predicting bipolar disorder behaviors. Psychiatrists can capture subtle warnings of an outbreak in time just by analyzing the patients' smart phones. When patients are in the depressive episode, the GPS sensor on their phones will send out the signal that they are being inactive. They make and receive less phone calls, and the time they look at the screen increases. When they are in a manic episode, they tend to move around more, send more texts, and make more phone calls.

One research also finds that in comparison to regular Twitter users, users with schizophrenia post tweets about depression and anxiety at a higher frequency, which aligns with the clinical symptoms observed in labs. It also shows that the more one engages with social media and entertainment apps, the less stress and irritative emotions one will have. This offers the possibility of using online platforms to draw phenotype data for mental disorder symptoms that can predict and manage the illnesses through a new approach.

Furthermore, human-computer interaction can be useful for people unaffected by mental disorders in predicting their emotions. In another study, scientists used smart phone sensors to predict the emotional changes for 32 healthy participants within two months. The items that are tracked over this period include the number and length of phone calls, text messages, and emails, number and mode of smart phone apps, memo of visited websites, and locations. The accuracy rate of this study for predicting emotional changes has reached 66%, and it can be raised to 93% when personalized prediction model is applied.

Disease assessment is another aspect that phenotype data can be useful in. Aside from doctors' subjective biases, the clinical interviews and assessments conducted in the traditional diagnostic phase are also subject to other limitations. First, these assessments are un-ecological, which means that they are conducted when participants are detached from activities in their regular lives. Second, these assessments can be occasional. The limiting factors such as location and assessment personnel make it hard for these assessments to be expanded.

By tracking the patients' experiences, actions, and emotions in their everyday lives through portable electronic devices, phenotype data provides long-term, sustainable assessments that alleviate the problems in the traditional method. It enables more time and opportunities for doctors to learn about their patients and propose new treatments that can reduce the likelihood of their disorder outburst.

Finally, with disease treatment, information acquired by wearable and mobile devices and social media is an essential supplement to the traditional assessments. A case study of a neurology phenotype data that tracks the patients within a certain community suggests lithium salts proves ineffective in soothing the symptoms of amyotrophic lateral sclerosis. Later, this finding was replicated in several more expansive and more time-consuming randomized controlled trials. This phenotype data also facilitated in modulating and personalizing the treatment plans.

In addition to the positive readings of phenotype data in medication, there are some challenges that needs to be acknowledged as well.

First, uploading medical information to public apps entails risks for both the patients and clinical support workers. Theoretically, privacy laws prohibit sharing of mental health-related data. The Health Insurance Portability and Accountability Act (HIPAA) that has been enacted for 24 years in the US has a clear regulation on medical data sharing, and the General Data Protection Regulation (GDRR) in Europe is also against this action. Nevertheless, in a report published by Privacy International in 2019, it shows that some popular websites on depression in France, Germany, and the UK have been violating this rule by disclosing user information, test results, and test answers to ad businesses, data managers, big tech companies, and other third parties.

Second, ethicists worry that phenotype data is blurring the boundaries of medical data categorization, management, and protection. If we view the details of our daily lives as proof of our mental health, then every random stranger can access our mental condition through the digital records. Everything about the way we live—the words we say, how fast we respond to each text message or phone call, how often do we browse the Internet, which posts we like—will be public. We cannot hide from the traces we create and leave behind.

Dr. Nicole Martinez-Martin from Stanford University said, "This technology has brought us somewhere outside the traditional mode of protecting certain kinds of information. When all data has the possibility of being healthy data, it'll be a huge issue whether healthy data exceptionalism is still meaningful."

Third, phenotype data acquired through smart phones and other wearables must prove its clinical effectiveness. It is still uncertain whether data-based decision-making and improved efficiency are beneficial to lowering the incidence rate, recurrent rate, and death rate. It is rare to expect medical research depending solely on monitoring to provide satisfying clinical results. In addition, existing studies are mostly done in constructed environments or labs with mentally healthy participants. The subject and time for these studies are both highly restricted.

With that said, phenotype data represents a new powerful instrument in administering future diagnoses for mental disorders, and the present-stage objective for it is to reduce harm both on a scientific and individual level and to increase clinical efficiency that could bring concrete benefits on a large scale.

- **AI and Visual Recognition**

The COVID-19 pandemic has been a critical impetus to AI application in the medical field. Relevant technologies prove to be efficient in fighting the war against the pandemic, making it a litmus test for AI's potential value in the healthcare industry. But the development of AI application infrastructures is still rudimentary, especially in the areas of visual recognition, remote diagnosis, and healthcare management.

The sub-category of visual recognition, which serves as a diagnosis assistant, has the most diverse application scenarios in medication. Originating from oncology, the denotation of medical imageology has gradually expanded to other areas of study. Conventionally, the process of medical imageology pretreatment and diagnosis would require four to five doctors. However, with the help of AI visual recognition that trains computers to analyze medical images, only one doctor is needed for quality control and final confirmation. This will be a fundamental benefit for improving medical practice efficiency.

In comparison to statistics collection (e.g., medical history) that takes years to complete, imaging data is both easy to acquire and to access. Each image takes only a few seconds to produce. It reflects most of the information that doctors should know about the patient and can be used as direct evidence for making treatment plans.

The massive quantity of imaging data, the relatively reliable data foundation it provides, as well as the development in visual recognition algorithms are making AI application in medical imageology studies possible.

The technologies of visual recognition and Deep Learning lie at the heart of medical imageology. In a clinical diagnosis, visual recognition will be applied first during the perception process. Then it will process and analyze the unstructured image data and extract useful information.

Then, DL technology will be used to insert a huge amount of image data and diagnostic records into an AI model and conduct DL training with a neural network.

Finally, the continuously improving algorithm model will provide personalized diagnostic results based on intelligent inferencing of image data.

This AI technology in medical imageology mainly addresses three needs. First, lesion recognition and labeling. AI medical imageology products will perform segmentation, feature extraction, quantitative analysis, and comparative analysis. There are multiple lesion recognition and labeling systems such as X-ray, CT, and NMRI. Current AI medical imageology systems can process more than 100,000 images within seconds. It helps to increase diagnostic accuracy, especially by alleviating false negative occurrences.

Second, automatic target delineation and Adaptive Radio Therapy (ART). These products can help radiation therapists delineate a set of 200 to 450 CT scans within 30 minutes. What's more, throughout the 15 to 20 times that a patient receives the scan, the AI medical imageology technology can continue to identify changes of lesion location and achieve adaptive radiation treatment. This will effectively minimize radiation's damage on the patient's healthy body parts.

Third, image-based 3D reconstruction. The timesaving Gray statistic registrations and feature-based registrations will make a difference in locating lesions, detecting extensions, differentiating between benign and malignancies, and designing operation plans.

Right now, AI medical imageology products in China mainly focus on tumor and chronic disease screening in the thorax, head, pelvis, and limbs.

Pulmonary nodule (PN) and diabetic eye screening were the hotspot areas in AI medical imageology in its rudimentary stage. In recent years, breast cancer, cerebral stroke, and bone age detection are also gaining popularity.

In addition to image data's accessibility, policy and capital support is another important reason for AI technology's continuous growth in medical imageology.

Between 2013 and 2017, multiple policies have been released on supporting domestic products, third-party diagnostic centers, and studies of telemedicine.

For example, by the end of 2016, the State Council of the PRC issued the "Thirteenth Five-Year Plan: National New Strategic Products Development Plan." The document highlights developing high-quality medical imageology devices and supporting corporations, healthcare facilities, and research institutions in building third-party imageology centers. In January 2017, the National Development and Reform Commission included medical imageology products in "Content of New Strategic Products, Priority Products, and Service Guidance."

On November 15, 2017, the Ministry of Science and Technology held a launch meeting in Beijing for a new generation AI development plan and major technology programs. During this event, Tencent Miying was recognized as an open, innovative platform for new generation AI medical imageology technology. It is worth mentioning that Tencent Miying and Tencent Cloud CT devices have been implemented in several hospitals in Hubei during the COVID-19 pandemic and proved helpful in diagnosing cases for medical staff.

In terms of capital support, Global Market Insight data report shows that AI medical imageology owns the second largest market in the healthcare industry and a growth rate at above 40%. By 2024, it will reach US$25 hundred million and take up 25% of the market size.

Empowered by accumulating data, maturing algorithms, and existing experiences, the commercial model for AI medical imageology development is becoming clearer. The two prevailing models are: 1) cooperate with primary hospitals, private hospitals, and third-party test centers, provide diagnostic imaging services (DIS), and charge-by times; and 2) cooperate with general hospitals, test centers, third-party medical imageology centers, and medical device manufacturers, provide technical support and solutions, and accept a one-time payment or payment by installment.

Over 100 corporations in China have already employed AI products in the healthcare industry, and most of their projects involve medical imageology. This frequency is much higher than other application scenarios. EqualOcean's report "AI Business Landing in China, 2018" shows that 10% of AI-related, non-listed enterprises in China are among the national top 100, and 60% of them are in medical imageology.

Competitors in this field are numerous. There are traditional medical device corporations such as GE Healthcare and Lepu Medical, Tech Giants such as Google, IBM, Alibaba, and Tencent, and startup companies such as YITU Technology, Deepwise, Shukun Technology, and Infervision. These market players display distinct strengths in capital support, market expansion, product designs, technology R&D, and many other aspects.

Although there hasn't been a monopoly enterprise in the industry, disparity is starting to emerge among the competitors. Starting from 2017, startup companies that focus on multiple types of disease and technical directions have attracted considerable funding. Some leading enterprises have completed a Series C funding round and are proceeding to technology and experiences transfer, disease and product pipeline expansion, and global reach with an aim to consolidate their advantages.

While technology advancement and support in policy and capital are providing an immense opportunity for AI medical imageology development, these factors also impose the following restrictions.

First, the healthcare industry pertains to life. Therefore, a false negative diagnosis is critical to AI medical imageology. Even 1% missed diagnosis can cause fatal consequences, and doctors will need to re-examine all screenings if this happens. Only when a zero false negative diagnosis is guaranteed can AI medical imageology provide real help for doctors.

Second, AI medical imageology is not at present an inevitable technology, so hospitals do not feel strongly inclined to have it without the demand from patients and a well-established insurance system to support their demand.

Third, although some enterprises have employed AI medical imageology technology for business purposes, the technology hasn't been centrally commercialized. Nevertheless, AI admittedly demonstrates new application modes that can give birth to businesses through deep integration.

As a central force in the new generation industrial reform and infrastructure construction, AI will bring fundamental change to the traditional medical institution's operation and ameliorate its pressure on resources. In the post-pandemic era, AI in medication is likely to welcome major development, especially in diagnostic radiology practices.

3.2 AI in Finance

Fostering economic society's digital transformation has now become a global consensus thanks to the booming of scientific and technological innovations. The most heated discourse in year 2020 concentrated on finance digitalization. This concept generated new business models, applications, procedures, products, and client relationships, leaving a profound impact on financial institutions, markets, and services.

AI plays an indispensable role in the finance digitalization process as a critical driving force in the Digital Revolution and industrial reform. It can be anticipated that the future of financial technology (FinTech) will highly depend on computers replacing and surpassing part of human management experience and competence.

• Revolution in Financial Production Brought by AI Finance

At present, AI technology can be found in every aspect of our lives, and it has a natural coupling with finance. The special nature of the financial sector poses new challenges for AI development, and the breakthroughs in AI finance will be beneficial for seizing further opportunities and occupying commanding points in technology. The integration of AI and finance is a definite trend. With specific technologies such as big data, cloud computing, and blockchain, the future of the financial industry will be revolutionary with endless possibilities.

In the case of China, AI financial development is helpful in enhancing the finance sector's adaptability, competitiveness, inclusiveness, and a financial institution's capability in identifying and preventing risks. It will stimulate supply-side reform, facilitate real economy, and improve people's living skills.

In comparison to Internet finance and FinTech, AI finance has a tremendous advantage in financial productivity. Although AI also relies heavily on big data and cloud computing, it applies sensors and DL algorithms instead of in-depth data mining throughout its operation, with the aim to mimic human sensory systems and gradually take over the intense data process tasks.

Currently, the foreground application scenarios of AI focus on changing financial service enterprise acquisition and preserving customers, which will provide more opportunities for market innovation, in the forms of intelligent marketing, AI customer service, and robo-advisors.

Robo-advisors refers to financial models built on an AI algorithm and Modern Portfolio Theory and tailored to the investor's risk preference, financial situation, and revenue target. It will offer personalized asset allocation advice based on continuous tracking and dynamic re-equilibrium adaptation. So far, there have been pilot projects on robo-advisors, but many questions remain to be explored before it can be fully promoted nationwide.

In comparison to traditional human investment advisors, robo-advisors provide an efficient and extensive counseling service, with low investment threshold, reduced fee rates, and high transparency. It is unaffected by human subjectivity and can thus achieve high-level investment objectivity and diversification. In addition, its financial management service and scenarios are rich and customized.

AI is not only applicable to the work in the foreground but also that in the middle office and backstage operations. Some enterprises have been optimizing AI algorithms to obtain more precise investment projections, higher profits, and lower tail risks. In brief, robo-advisors have a huge potential in its future profitability by gaining excess proceeds through combinatorial optimization in firm offers.

Smart credit assessment is notable for real-time operations, automatic judgement, and short review cycles. It is especially helpful for micro-credit companies in transforming to a more efficient service mode and has already been put into use in E-banks. Smart risk control is applied in banking enterprise loans, online financial loans, credit assessment in consumption scenarios, risk pricing, and receivables recovering, providing a new type of online-service-based risk control model for the financial sector.

Although AI finance is still in its primary stage concentrating on intelligent transformation of procedural and repetitive tasks, it is certain that AI technology will make possible the complete automation of client's financial life as AI applications gradually move from the peripheral services to the core businesses.

• Risks and Challenges

Aside from the amazing opportunities brought by the progress from human service to computer service, AI finance also entails new challenges against existing regulations, ethics, and disciplines.

For instance, a data quality or algorithm defect may cause an investment loss. Due to the fundamental role of computation in AI finance, procedural errors such as over-fitting can enact the butterfly effect and pose systematic risks.

The long tail effect and network effect enable better client solicitation and risk control performances at a lower cost. However, they also increase the complexity of the financial system and can possibly make the risks more contagious and consequential. This will then trigger the bandwagon effect and enlarge the impact of finance procyclicality.

Also, with intelligent processing of big data lying at the heart of financial policymaking, the risk of a data breach regarding personal or sensitive information accentuates the demand for privacy protection and data security. At the same time, a non-transparent algorithm and data that is incomplete, unrepresentative, and defective will affect the decision-making processes.

Therefore, it is crucial that financial institutions understand the AI systems as well as all possible negative circumstances that clients may come across. They should also take responsibility for the bias caused by a flawed algorithm.

In order to cope with these problems, there needs to be a multidimensional cooperation system involving the three parties of government, markets, and society.

On the one hand, AI finance needs to follow the general rules in AI management, but it also needs to take the uniqueness of the financial sector into consideration and place dual emphasis on innovative application and risk prevention. To achieve this goal, we need to support innovations in AI finance industrial models and implement effective supervision measures.

From 2019, policies of the People's Bank of China and China Banking and Insurance Regulatory Commission (CBIRC) regarding AI finance have mainly addressed the issues of regulation tightening, technical support, and loan services for micro-credit businesses. In recent years, financial services have seen a trend of deepening connection with the internet, increasing complexity of business scenarios, and downplaying of boundaries. The thriving of AI finance poses a challenge for financial management. After the P2P scandal in 2018, the supervisory authority imposed greater forces in financial regulation.

For innovation support, in January 2020, the People's Bank of China released "Announcement on Pilot Applications of Innovative Supervision on Finance Technology (Pilot Test I, 2020)" which demonstrated a flexible supervision experiment for AI, blockchain, and open banking API in a simulating scenario through the model of a Supervisory Sandbox. According to the announcement, the experiment showed positive results in reducing operational risks and technical uncertainty, and trial and error proves to be a better choice in exploring FinTech supervision.

In the future, technology enterprises will disengage themselves from financial service businesses, focus more on technology output, and propel market supervision.

On the other hand, many organizations, such as the privacy committee of the Bank of America, are starting to investigate strategies in response to the ethical questions of AI finance. Google suggests applying a person-centered method, evaluating and supervising through various indicators, and widely inspecting data availability when checking the deviation sources. The Department of Finance of Canada also issued a guidance document on AI quality, transparency, and public accountability.

Rational development and utilization of AI finance require explicit guiding principles and security methods, including the impartiality, interpretability, and robustness of algorithms.

Institutions which utilize AI finance applications must take responsibility for data processing and make sure that research staff who develop, verify, and supervise AI algorithms are qualified and experienced. They should be able to identify possible social biases and historical aberrations and methods for correcting them. The institutions also need to have internal policies and management mechanisms that are updated regularly so that algorithm surveillance and transparent risk control system can be ensured.

The future of financial services lies in the adequate application of AI technologies, which is a long-term, upward spiraling process that asks for multilateral collaboration among enterprises and various political and social organizations. Only in this way can the benefits of new technologies be maximized.

3.3 AI in Manufacturing

An industry with strong vitality and creativity, manufacturing has a pivotal role in assuring sustained economic prosperity and social stability.

Industrial development empowers human civilizations in transforming natural environments and acquiring resources. Its outcome is directly or indirectly utilized in social consumption activities and greatly improves people's living standards. In a way, industrial manufacturing has been defining the existence of human society ever since the Industrial Revolution.

However, in recent years, the downside of industrialization began to be exacerbated. The deindustrialization in developed countries caused high unemployment rates and trade deficits, while low value industries in developing countries resulted in deterioration of profits and the industrial environment. With a huge number of manufacturing enterprises on the edge of existential crisis, the need for manufacturing digitalization, networking, and intellectual transformation is critical and urgent.

The rapid advancement and extensive use of AI technology are drawing manufacturers' attention to AI manufacturing. But there are still many obstacles that need to be overcome before this new concept can turn into profitable products.

• Challenges in AI Manufacturing

AI technology can help optimize efficiency in every stage of the manufacturing process, collect production data through the Industrial Internet of things (IIoT), and achieve autonomous optimization through DL. Right now, major directions of AI manufacturing include intelligent product development, quality and efficiency improvement, and supply chain intelligentization.

During the manufacturing process, AI technology can search for all kinds of possible plans according to the requirements of the target products, come up with designs through smart production, and present the result in an integrated, commercialized form. Smart phones, industrial robots, service robots, self-driving cars, and drones are examples of new generation smart products. These machines do not simply carry out mechanical tasks but can operate automatically under many complicated situations.

An intelligent supply chain requires high-quality demand prediction, which can efficiently help enterprises to adjust manufacturing plans and improve factory utilization rates. In addition, an Automated Guided Vehicle (AGV) can help achieve automatic storage optimization and save labor costs.

Manufacturing digitalization is the foundation for all the stages mentioned above. However, the informatization level of manufacturing businesses in China is uneven. Due to the sophistication of the manufacturing industry chains, which entails professional knowledge of the enabler, the development of AI manufacturing is much harder than other industries.

The manufacturing industry is huge and historically marked by intricacy and fragmentation. A factory is usually found with production equipment from several manufacturers, following different technical and data standards, but with no common interface. Product variation is more prominent if they are from different factories or different manufacturers. This makes traditional manufacturing informatization more challenging.

Before 2018, the digital industrial economy only took up 18.3% of the overall economic output in China. The general downturn of the digitalized economic environment implies a low penetration of AI technology in manufacturing.

In general, the value of AI manufacturing is hard to measure in the current stage. Some enterprises, especially small and medium enterprises, are not sufficiently motivated to apply AI technologies. While AI technologies mainly aim for optimizing brands, improving product competitiveness, increasing profit, or reducing operation costs, small and medium enterprises that prioritize survival instead of expansion usually find these goals irrelevant to their business models. In other words, small and medium enterprises cannot afford the heavy cost pressure of AI manufacturing despite its appealing reward.

Even large enterprises struggle to visualize AI applications in some industrial sub-sectors with a clear understanding of the potential risks, profit, costs, and corresponding plans. The intelligentization process will be extended and arduous given the long-term overproduction and excessive data.

• Humans as the Core of Intelligent Production

There is a fundamental difference between manufacturing intelligentization and automatization. Intelligent production doesn't mean it is automatic or unmanned. To attain intelligentization is the key to untangle the knots in AI manufacturing and make industrial upgrading a reality.

Automatization means machine automatic production, and the ultimate pursuit of this process is machinery replacing humans in a large scale. Intelligentization, on the other hand, refers to machine flexible production and emphasizes human-machinery cooperation. Machines should be able to accommodate the needs of human workers and adjust to parameter changes.

In this light, AI manufacturing aims for reconstructing people-oriented production modes that assign to machines the simple, repetitive, and sometimes dangerous tasks on the tremendously refined assembly line, while human workers take on more of the innovative and management tasks.

In order to achieve cooperative intelligent manufacturing, machinery needs to be versatile with more intelligent functions. A good example is the industrial robot that connects individual production devices into a whole production network and depicts the application scenario for a fourth industrial revolution. The outlook of AI manufacturing in the twenty years to come is groundbreaking, with every single industrial robot in the global workplace connected to each other and the traditional understanding of production overturned.

It is necessary that industrial robots are capable of interacting with humans under complex and uncharacterized circumstances. Also, more intelligent platforms need to be built to expand AI applications in more industrial scenarios and make comprehensive arrangements.

The basic conditions required for AI manufacturing are already acknowledged in the *2020 AI and Manufacturing Integrated Development White Papers* issued by the US National Industrial Security Institute. Nevertheless, single-point AI devices alone are not enough for a business management upgrade. We need to place more attention on the entire value chain.

3.4 AI in Retailing

Since the term "new retailing" was coined in 2016, new programs sprang up like mushrooms within just a few years. Internet giants led by China's Alibaba and Tencent made huge investments in physical business management and created many new brands, such as Fresh Hema (Alibaba), 7fresh (JD), Zhangyu Shengxian (Meituan), Super Species (Yonghui), etc.

The investment craze on new retailing reached a climax in 2017 but soon began to cool down considerably in 2018. Many unmanned retail businesses collapsed during this time. In 2019, the market began to reflect more rationally on the different business models for new retailing.

New opportunities arose in 2020 as live commerce quickly went viral during the pandemic. The rapid development of a series of cloud industries propelled the penetration of Internet and digitalization in the traditional industries and injected fresh momentum into the new retailing businesses.

- **The History of Retailing**

Before the 1990s, the retailing businesses in China were mainly in the form of specialty stores. Only later were they reorganized following a changing society's demand and turned into department stores.

During the 90s, supermarket chains dominated the retailing market. There were only a few modern specialty stores, hypermarkets, and convenience stores at this time, and they were at much smaller scales than the ones in other countries. For example, in 2000, the volume of the biggest supermarket in China, Hualian, was only 1/80th of the contemporary Walmart. The intense competition among supermarket chains forced the market into an acclimatization stage.

Around this time, GMS and discount stores appeared, and foreign retail businesses like Carrefour began to show up. It was a new era for the Chinese retailing market. From the year 2000, the number of GMS stores in China proliferated. Development of shopping centers with entertainment facilities, restaurants, shops, and recreation areas all-in-one made the outlook of retailing businesses highly promising.

Yet, Internet and e-commerce expansion was posing a growing threat to undermine this vision. Approaching the year 2013, people's consumption awareness and habits changed significantly, giving rise to an online shopping fever while the offline stores suffered from a severe depression.

Then, the focus of e-commerce began shifting from PC to MT. In 2015, e-commerce reached a phase of stable development, and many physical retail enterprises started looking for ways to cooperate with e-tailers in the forms of Internet plus and (Online to Offline) O2O.

These endeavors turned out to be desperately needed, because the Chinese retailing market underwent another major repercussion in 2016. Offline supermarkets closed down one after another, with RT-MART the most shocking. At the same time, the situation of e-commerce was looking less positive than before as more competitors began to join in and share the profits. Therefore, on October 13, 2016, Ma Yun (Jack Ma, chair of Alibaba at that time) indicated in an Alibaba Cloud conference that "The age of exclusive e-commerce will soon be over. There will be no e-commerce in the next ten to twenty years, only new retailing."

So what is new retailing? Ma's explanation for it was "New business models that can only be realized when online and offline logistics are combined. It is a model that aims for improving traditional retailing efficiency through digital methods." In 2017, this proposal finally came into being. Fresh Hema, for example, was the result of Alibaba's reconstruction of offline supermarkets. It soon became the prototype for all Alibaba subsidiary businesses in new retailing. Customers can both shop in stores and make an order online. Fresh Hema is best known for its fast delivery service. Goods will be delivered to customers living within three km from the store in 30 minutes.

In addition, Alibaba also bought into the appliance digital linkage Suning, Intime Department Store, SanJiang, Sun Art Retail, and cooperated with Bailian Group and Starbucks.

- **New Retailing as a Service Revolution**

The development of new retailing is both driven by external factors such as consumption and technical upgrades as well as internal factors such as transformation within the industry.

The changing patterns in economic growth, personal income variations, and demographic shifts are transforming the main body of consumption in China. Wealth structure indicates that despite the negative impact of trade friction and the relatively mild growth rate of average disposable income, the medium number has been constantly rising. Strategies regarding poverty alleviation and inter-regional cooperation thus proved effective in improving income equality, and lower-middle-income population's need in higher marginal consumption can be better satisfied.

Based on demographic structures, we see that the "post-90" and "post-00" generations are now entering a phase where they have better control over various aspects of their lives, and are starting or about to start an independent living. These two generations, numbered around 335 million in China, grew up in an era of a thriving economy. They possess stronger inclinations towards marginal consumption, greater capacity in autonomous consumption, higher need for high-quality diversified consumption, and value the content and services the retailer provides more.

The new round of information surge provides necessary support for retailing businesses. Cloud computation, big data, IOT, AI, VR/AR, and other new-generation technologies have now become a major impetus in generating innovations in retailing facility constructions (flow rate, logistics, payment, real estate), making them flexible, intelligent, and cooperative.

Viewed from a holistic perspective, digital technology empowers every stage in the retailing industry chain: production, logistics, and sales. First, commercial digitalization can considerably enhance product accessibility and encourage more consumers to engage in the product design process. Second, consumption digitalization (e.g., VR/AR) can facilitate information exchanges between the retailer and the customers and make sure the latter's demands are identifiable, accessible, discernable, and servable. Third, post-consumption digitalization allows customers to be fully aware of their purchasing activity and subsequent logistic information.

The internal transformation of retailing businesses derives from the dual demand of traditional retailers and e-tailers. As mentioned above, e-commerce imposed intense pressure on physical entities. By March 2020, Chinese online shoppers had reached 710 million. The total transaction size in 2019 grew by 16.5% from the previous year and reached 10.63 trillion RMB.

Online retail sales reached 10.63 trillion RMB, of which online retail sales of physical goods reached 8.52 trillion RMB, accounting for 20.7% of the total retail sales of consumer goods. In the meantime, the transaction size of physical retailing was continuously decreasing and desperately in need of new opportunities. New retailing, of course, was exactly what it was looking for.

As for exclusive e-commerce businesses, it is gradually losing its advantage under low competition pressure. Until June 2019, in China there were already 639 million online shoppers among 854 million netizens. In other words, 74.82% of Chinese netizens shopped online. It is difficult for this number to keep growing at a rapid speed, and it is difficult for e-commerce businesses to attract people with limited purchasing power (e.g., people from underdeveloped regions, older people). According to JD's annual report, its customer acquisition cost has risen from 80 RMB to above 1,500 RMB within the five years between 2013 and 2018.

As many e-commerce businesses were eliminated during the intensified competition, the ones that are left became more stabilized and mature in their operations. They also realized that a low price strategy is getting obsolete in the present market structure. Product, service, experience, and data will occupy the vantage point in transactions. In brief, new retailing is all about providing the ideal service for customers throughout the consumption activity. It is a service revolution.

• New Retailing Supported by AI

New retailing digitalization is inevitable. AI is now visible in all stages of the retail value chain, and AI applications in retailing businesses are beginning to converge.

First, AI new retailing can provide individualized recommendations for customers based on their purchases and search records. It thus enables retailers to quickly adapt their sales strategies. AI custom services can provide help 24/7, which saves a great amount of labor costs. Computer vision allows transactions to happen without contact. In general, AI helps retailing businesses to better anticipate customer's needs, set prices, display products, and make sure customers connect with the appropriate goods at the appropriate time and location.

Second, AI helps retailing businesses achieve higher proficiency in supply chain management. With computer vision, retailers can identify products, inspect damages, and protect payments at a lower cost and higher speed. One major challenge that traditional retailers face is maintaining accurate inventory. But AI can solve this problem by connecting all the links in the supply chain and consumption chain, keeping retailers updated with detailed information about stores, customers, and products, so that they can make better decisions on inventory management. In addition, AI can promptly identify inventory shortages or misplacement and inform human workers.

Third, AI new retailing will redefine the context of retailing based on customer experiences. Labor productivity (sales divided by staff numbers) in retailing businesses began to drop since 2018. But AI technologies such as computer vision and NLP will effectively address this deficiency by providing high-efficient, low-cost customer service. In the foreseeable future, new retailing will take place in highly contextualized and personalized scenarios.

According to iResearch, China's total investment in retail digitalization through modern trade in 2018 was 28.51 billion RMB; among which, 90 million (3.15%) were in AI. It is anticipated that this investment will exceed 70 billion RMB in 2022 with 17.8 billion RMB (25%) in AI. Thanks to the leadership of retail giants like Alibaba, JD, and Suning, AI new retailing will continue to grow with high momentum.

At the same time, it is important to acknowledge some challenges this new industry will face in the near future.

First, online platforms and offline stores vary considerably in operation systems, marketing strategies, and product arrangement. The ideal form of cooperation between the two entities is to have online platforms introduce customers to offline stores where they are served. But conflict about inter-trade products, prices, and logistics may arise during the integrating process.

Second, high operational costs may lead to difficulty in gaining profits. New retailing requires a huge amount of investment in human labor, capital, and material. This is without doubt a heavy burden for the new industry. Right now, most new retailing projects are still in an unprofitable state, and many enterprises collapsed due to intense competition as well as trade and store expansion costs. Therefore, many offline retailing businesses are subject to financial pressures and restrained by existing, low-risk operational models.

Third, new retailing also needs to deal with operational problems generated by regional expansion and replication. For many new retail stores, expansion is the only way to acquire more customers. However, rapid expansion brings challenges to staff and product management. For example, Fresh Hema has been accused starting from 2019 of having a regional bias, misplacing kitchen waste, and selling expired food. For new retailing businesses, there is still a long way to go before they achieve supply chain upgrading and reinforce product quality control, staff training, and regulation implementation.

The world changed considerably in 2020, and the development of the digital economy was part of the change. With the landing of AI technologies in new retailing, the industry is getting smart and cooperative both in the supply end and the consumption end.

As an important component of the digital economy, the new retailing industry is capable to push some boundaries and influence the development of more domains. Already, new retailing is not monopolized by e-commerce businesses. Logistics, real estate, health care, and many other professions are all adopting the operational mode of new retailing to gain in-depth empowerment. This is a sign for new retailing to enter a new phase of development. Soon, there will be more new retailing-based models integrated in more businesses.

3.5 AI in Agriculture

Agriculture has been the foundation of human existence and national economies since ancient times. But a booming population and urbanization progress are posing severe challenges to it since the area of arable land is decreasing fast. The world has been searching for ways to alleviate the pressure through information technologies; among which, AI-based intelligent modality proved to be most fruitful.

Starting from the reform and opening-up period in 1978, China's level of agricultural development has been greatly enhanced. But the downside of this trend is gradually being revealed as well. For example, land resources are running low, agriculture industrialization is inadequate, agricultural product quality is insecure, and ecological environment is destabilized. How to achieve sustainable agricultural development in the case of resource shortages is a major challenge that China needs to tackle on the road to industrialization, informationalization, urbanization, and agriculture modernization. Right now, using AI technologies to increase productivity draws the most attention in agricultural research and application.

• From Extensive Agriculture to Precision Agriculture

Agriculture, in the broader sense, implies farming, forestry, animal husbandry, fishery, and agricultural auxiliary activities. So far, AI implementation has been mainly manifested in three areas: precision farming, precision breeding, and protected agriculture.

Precision farming is an interdisciplinary, comprehensive technology developed in the 1980s. It features the "3S" (e.g., RS, GIS, GPS) and the comprehensive application of automation technologies that optimize the control of agricultural activity. Precision agriculture addresses specific requirements of each field operating unit and regulates its investment on the crops, based on the knowledge of soil properties in the field, special variations of productivity, and crop production objectives. The aim of precision agriculture is to achieve a systematic diagnosis, optimal formulation, technical assembly, and scientific management. It guarantees the same or higher profit with the least amount of investment. As integration of new generation information technologies such as AI, IoT, 3S, and big data in agricultural production becomes prevalent, precision agriculture is highly reputed for its rational utilization of resources and positive impact on increasing crop yields, saving production costs, and improving the ecological environment.

Precision breeding refers to the innovative, Internet-platform-oriented husbandry model based on smart devices for precise feeding, automatic disease diagnosis, and automatic waste recycling. For example, large chicken farms with automatic production lines, automatic fecal cleaning systems, and smart egg collecting and sorting systems. In addition, Dr. Li Daoliang and his team from China Agricultural University also invented an aquaculture monitoring and management system that is both efficient and environmentally-friendly. This precision feeding system can track water quality, receive alarms, and adjust control equipment based on the results. Breeding personnel can also check on this information timely through smart phones, PAD, and computers.

Currently, the application types in precision breeding are limited, and they mainly apply visual and voice recognition in increasing animal survival rate and product quality. In comparison, higher accomplishments have been achieved in precision farming. AI technologies (i.e., satellite remote sensing, intelligent agricultural machinery, agricultural IoT) are making a difference in various aspects of farming activity, including sowing, fertilization, irrigation, weeding, pest control, picking, and sorting.

Protected agriculture is a recent concept of the agricultural industry with high levels of intensification and an important component of modern agriculture. With IoT, it can collect important data in the greenhouse, such as air temperature and humidity, soil temperature and humidity, carbon dioxide concentration, light intensity, etc., and initiate intelligent management of greenhouse planting through computers and smart phones.

The study of an artificial light plant factory started in 1974 in Japan. By the end of 2016, Japan owned 254 plant factories, ranking first in the world in terms of number, space, and output. The daily output of the biggest plant factory Spread, for example, is around 25,000 heads of lettuce, which makes its yearly output nine million heads. The facility agriculture industry in Japan is best known for the following features: large-scale greenhouse construction, indoor technology integration, diversification of product types, operation technology mechanization, industrialization of production technology, diversification of covering materials, soilless cultivation techniques, and biological control technology.

Plant factories are now playing an important part in modern agriculture development, especially in developed regions. It helps to alleviate the tension between the world's growing need for resources and the endurance of an ecological environment; it also helps to solve the problems related to food quantity, quality, and security. It is now considered the fourth type of agriculture succeeding land cultivation, protected horticulture, and hydroponic cultivation.

- **Constructing an Industry Information Integrated Service Platform**

An industry information integrated service platform is crucial to AI agriculture.

The integration of big data and the whole agricultural industry chain and the construction of agricultural information service platforms enable intelligent decision-making before, during, and after agricultural production. For example, Deep Learning systems can automatically diagnose crop diseases and insect pests through visual recognition. An atlas of agricultural technology knowledge can provide intelligent Q & A services, such as real time announcement of agricultural product prices, supply and demand analysis, and intelligent matching transactions. AI will assist and replace humans in the decision-making process to a considerable degree and achieve intellectualization in agricultural information services and precise voice delivery services. Other decision-making models like core crop item databases, knowledge maps, and crop growth models, when connected with IoT, UAV data, and satellite data, offer information guidance for the whole industry chain.

Also, e-commerce platforms and online marketing can maximize the benefit of O2O to efficiently process information resources, reduce agricultural production costs, and improve the relationship between suppliers and farmers. These facilities can build an omnidirectional management network that covers the entire process of agricultural production and circulation and ensures product safety in every link. Basically, the goal is to realize the brand effect of traceable agricultural products "from the field to the table."

Like many AI industries, the development of AI agriculture is still in its primary stage with many problems waiting to be solved. The future of AI agriculture calls for more in-depth study in acquiring information through massive agricultural data, summarizing patterns, trends, and correlations, and predicting possible application prospects for agricultural production.

3.6 AI in Cities

In today's world, building intelligent cities is no longer an uncertain or debatable concept but the megatrend for city development in the midst of the digital revolution. How to grasp the opportunities brought by the intellectualization trend is a common proposition that interests all cities.

An intelligent city is a comprehensive carrier for the final landing of artificial intelligence application scenarios. Cutting-edge technologies upgrade city infrastructures and bring intelligent city building to a higher level in all aspects. At the same time, problems of rapid urbanization, including dense population, single energy structure, low efficiency of resource distribution, high risk of transportation and logistics, low waste-recycling rates, and poor air quality set higher requirements for the development of AI industry.

• An Intelligent City is Not Just an AI City

Intelligence is usually perceived as a trait possessed exclusively by creatures (humans) with vital signs and various physical senses. Therefore, the term intelligence contains the connotation of life. Intelligent cities can be understood as living cities, and the definition keeps evolving with the continuous growth of those lives.

At first, intelligent cities are referred to as digital cities. As the concept developed within broader urban scopes and people's understanding of it deepened, they realized that the essence of intelligent cities is to attain sustainable urban development. It is to exploit resources more efficiently and provide a better life quality through emerging technologies, which have always been closely related with city expansion. With the help of the Internet, IoT, cloud calculation, and big data, cities are growing from static to dynamic, from imaginary and visible cities to Ron Herron's "Walking Cities." These features that reflect modern technologies are contained in the concept of the intelligent city.

The core technology of intelligent cities is Smart Computing. It can connect different industries, such as urban management, education, healthcare, transportation, and public utility. As the technological source of intelligent cities, Smart Computing influences urban operation from various aspects (e.g., municipal administration, construction, transportation, energy, environment, and services).

Although the academic circles have different emphases on the definition of smart city, the "six dimensions of intelligent cities" raised by Dr. Rudoph Giffinger from the Vienna University of Technology in 2007 has received general acknowledgement. The six dimensions are:

- smart economy—innovative spirit, entrepreneurial spirit, economic images and trademarks, industrial efficiency, flexibility of the labor market, international network embeddedness, science and technology transformation ability.
- smart governance—decision-making involvement, public and social services, transparency of governance, political strategies, and perspectives.
- smart environment—reduce pollution to the natural environment, environmental protection, and sustainable resource management.
- smart human resources—education level, affinity for lifelong learning, social and ethnic diversity, flexibility, creativity, openness and participation in public life.

- smart mobility—local auxiliary functions, barrier free communication environment (between countries), availability of communication technology infrastructure, sustainability, innovation and safety, and transportation systems.
- smart life (quality of life)—cultural facilities, health status, personal safety, living quality, educational facilities, tourism attractions, and social harmony.

The six dimensions cover all fields of urban development. In addition to the material elements, the framework also emphasizes the roles of society and people; it also sets high quality of life and environmental sustainability as its ultimate goal. In brief, the key of making cities more intelligent is to use information and communication technology to create a better urban life and sustainable environment. The ways to achieve this goal include bolstering the economy, improving the environment, and strengthening urban governance.

• AI in Future City Construction

AI is one of the important technologies for building intelligent cities in the future. AI applications in infrastructure building and comprehensive carriers can address a series of challenges in city development and facilitate secure and convenient city lifestyles and low carbon emissions. Eventually, it is expected that a thorough intelligent urban infrastructure innovation system will be built to cover the whole city.

Basically, AI applications in city infrastructure system can be divided into the following categories.

Green Space System

The green space system is a smart water resource management system that responds to the impact of global climate change on urban water environment, saves circulation, adapts to the environment, and ensures safety. The system includes the implementation of sponge city and rain garden technologies, which build a smart urban pipe gallery and form a more mature sponge city operation and management model. This enables an organic combination of a green space system, a sewage and waste disposal system, and an energy system as the reflection of a circular economy.

Transportation System

In intelligent cities, traffic application scenarios will be diverse. Transportation development will be associated with AI development that centers the needs of people and the city and aims for high efficiency, energy, and sustainability.

Traditional traffic operation in cities was bogged down during the pandemic. But the information service represented by communication networks and artificial intelligence technology not only regulates the conventional urban traffic network service, but also highlights the potential of remote cooperation mode under the information background that keeps the city vital.

Through informatization and intelligent convenience service technology, China has successfully made up for the deficiency in protected traffic (e.g., stranded transit passengers, difficult commutes for medical staff, and inconvenient medical travel for residents in closed communities). A reallocated transportation force restored order and vitality to the service system and ensured smooth travel for an important population. In the meantime, high-speed and convenient communication network and appropriate software platform, in the forms of high-definition live broadcast, remote conference, and others, not only offered the conditions for people's timely return to work but also ensured the basic needs of urban prevention, control, and treatment.

Logistic System

The logistic system of automatic circulation aims at low cost, high efficiency, and safety. It solves the problem of "the last kilometer" of logistics through automatic and intelligent means that enhancing efficiency of logistics distribution, reducing costs, and significantly improving the street traffic environment. An example of a logistic system in cities is the wireless charging infrastructure installed in the Netherlands urban underground pipe network.

Although strict quarantine policies during the pandemic have restricted residents' travel, information technology effectively guaranteed people's demand for living materials. With e-shopping platforms and distribution heat maps, delivery services are fast and efficient.

In view of residents' demand for community public services and taking the pandemic situation as an opportunity, the government and community are accelerating the process of informatizing relevant public resources. The process involves transforming passive responses into active service, making the independent and divided platform into a comprehensive and sharing one. "Zero run" and "video office" are becoming the new normal. It highlights the necessity of improving emergency material reserve mechanisms, optimizing smart health service systems, building community self-organizations, and coping management modes in the future urban construction.

Circulation Systems

Circulation systems, or solid waste treatment cycle systems, aims at waste recycling, reduction, and harmless disposal. It relies on an automatic pneumatic system and waste classification center to reduce urban greenhouse gas emission and improve the utilization rate of recyclable resources.

At the specific operational level, there are smart waste sorting systems, anaerobic organic waste power generation systems, pneumatic conveying systems for dry and wet waste, and kitchen processing systems. These can better control and operate waste circulation when combined with intelligent waste information monitoring and management network.

Sustainable Construction

Sustainable energy systems aim at high efficiency, flexibility, and renewability. Specific facilities that help realize energy conservation and recycling are Distributed Energy Resources (DER), afterheat pumps, Combined Cold, Heat, and Power (CCHP), microgrids, and energy storage technologies. At the specific operational level, we can try to form an energy chain in line with the concept of the circular economy by combining technologies such as afterheat pumps and waste incineration power generation in the urban comprehensive carrier. Give priority to the supply of renewable energy such as solar energy and wind energy in the guide block, and use distributed energy to supply the comprehensive functional area.

• Before "Intelligent Cities"

The ultimate goal of smart city development is to build a more high-quality urban life and a more sustainable urban environment. China's smart city construction at this stage, however, is still imperfect.

When using information-based means to serve pandemic prevention and resumption of work, the biggest challenge faced by local governments comes from data, including insufficient data collection and poor circulation. For example, in some cities, complete urban medical resources, epidemic prevention materials, and enterprise production capacity data are missing. They can only report the numbers temporarily by traditional means. The problems such as information petrification and repeated collection are prominent, which prevent these cities in supporting the epidemic prevention command organ to allocate effectively.

When promoting the resumption of work, in the face of the pandemic prevention pressure brought by cross-regional personnel flow, local data cannot be circulated and recognized, and it is initially difficult to integrate with medical, public security, transportation, and other information. This brings challenges to accurately grasp the health status of foreign personnel.

Intelligent cities map the real society through data. Only when the data are complete and flow smoothly can they capture valuable information in massive data, so as to realize the development of smart empowerment. In the developmental practice with cities and departments as the main body, there is still much data with obvious regional and departmental boundaries, such as social security information, marriage information, personal real estate information, etc., which are stored in relatively exclusive local departmental databases. The lack of communication between these databases sets artificial obstacles to the realization of the due functions of intelligent cities.

In addition, the protection measures of personal information security need to be improved. Large amounts of data provides effective support for pandemic prevention and control, but it also poses severe challenges to personal information security. For example, when individuals upload information through the platform built by enterprises, there is a problem of how to avoid information leakage or being used by enterprises.

With the mitigation of the pandemic situation, whether a large amount of personal health information needs to be destroyed should be planned to avoid occupying too many information storage resources. Preventing user information leakage is not only the bottom line of intelligent city construction but also the biggest challenge faced by intelligent cities in the future. We should both strengthen institutional regulations as well as technical reliability when classifying and managing data for higher security.

The intelligent city has created unlimited possibilities for future city development. Up to now, an intelligent city is no longer just a "technical commitment" but a "privilege interface" between a human-centered digital society and the real world. It entails further discussion on the realization of technical ability, policy design, and application experience as well as on digital ethics, digital equity, and the standardization and improvement of digital literacy.

3.7 AI in Government

Every scientific and technological revolution plays an important role in the major transformation of human political civilization.

During the first industrial revolution, British society formed a quasi-bureaucratic organization based on instrumental rationality, and the corresponding government management ideology and organization form also became a worldwide template for early governance modernization.

The second industrial revolution produced a new dynamic system that promoted the formation of a specialized division of labor and assembly line production mode. Max Weber's bureaucracy has become the mainstream form of global government organizations.

The third industrial revolution, marked by computer and information communication technology, advanced the emergence of service-oriented economies and e-government. Self-adjustment in bureaucratic organizations was achieved through seamless government and new public management programs.

Now, with the deepening of the information technology revolution and the rapid iteration and penetration of emerging technologies, technologies represented by big data and artificial intelligence have pushed human society to the fourth industrial revolution. The speed and breadth of new technology development and its impact on the economy and society are unprecedented.

The most remarkable feature of the fourth industrial revolution is the development and diffusion of digital technology, which led to the integration of the boundaries of physics, mathematics and biology, and fundamentally changed the way people live, work, and communicate. Once again, it sets a profound impact on national governance and government reform. The construction of a digital government with the core characteristics of data-driven and digital governance has become the core issue of global government innovation.

Today, the construction of digital government, which started in the 1990s, has embarked on a key node again.

• Digital Government in a Digital Age

The establishment of a digital government is inseparable from the framework of the digital age.

Since the middle of the 20th century, the digital revolution has sprung up all over the world. In the past few decades, with the significant improvement of computing power and the decline of corresponding costs, digital technology has made great progress, and today it has formed an interdependent and interactive digital technology ecosystem, including the Internet of Things, 5G, cloud computing, big data, artificial intelligence, etc.

Obviously, each technology contains unlimited application possibilities, while the digital technology ecosystem multiplies this to infinity. A digital ecosystem promotes the transformation of the whole economy and society, that is, digital transformation.

In the context of social digital transformation, the pressure and challenges faced by the public sector, with the government as the core, are more prominent.

On the one hand, it is how the government public sector can better play its role in solving many new problems and challenges and establishing an inclusive, trustworthy, and sustainable digital society. Inclusiveness means that not only are digital resources accessible to all but also everyone can enjoy its benefits. Trustworthiness implies that the data environment is based on privacy, security, responsibility, transparency, and participation. Sustainability digital society refers to co-existence and co-development of the economy, society, and environment.

On the other hand, it is how the government should deal with the transformation of the digital economy and society, establish a digital government, and create greater public value. The digital transformation of government is a systematic project, which is not only a technological change but also an institutional change for process reengineering.

Digital government is a leading component in "Digital China" realization. It is of great significance for narrowing the digital divide, releasing digital dividends, supporting the development of the Party and the country, promoting balanced, inclusive, and sustainable economic development, and improving the modernization of a national governance system and governance capacity.

There are different ways for government to approach these challenges. Using digital technology for governance and introducing emerging technologies (e.g., big data, cloud computing, IoT, AI) to improve government capacity can lead to all-round "technical empowerment" for governance. In the social digital transformation stage, government can create greater public value for the society.

• AI Empowers Digital Government

AI has made outstanding achievements in establishing an efficient, digital government.

Driven by an urbanization strategy, China has become the country with the highest urbanization growth rate in the world. In 2018, China's urbanization level reached 60% and the urban population was about 730 million. It is expected that the urbanization rate will exceed 80% in 2050, and the scale of urban population will further expand. Such a large urban population will produce a great number of government affairs. Through the application of robot process automation (RPA) and AI technology, administrators can be liberated from fixed and repeated work and focus more on improving urban quality and optimizing residents' living environments. Facial recognition and natural language processing, for example, can enhance government service levels, provide convenient and fast services for enterprises and residents, and help intelligent decision-making.

Government service is one of the core digital government constructions, and it is also the fastest growing field on the road to artificial intelligence. For example, in the southern China city of Shenzhen, the Shenzhen Public Security Bureau will soon change their traditional "face-to-face" windows to online handling, and citizens' requests will be received through "face scanning." There will soon be a unified government information resource sharing system in the city, collecting 385 types of information resources and more than 3.8 billion pieces of data from 29 units, so as to provide data support for the comprehensive and intelligent government services. Hangzhou has also built an integrated intelligent e-government management system. This extranet integrates the business systems of digital urban management, planning systems, and financial systems and provides one-stop services.

At this stage, due to the separation of government departments and the great differences in the intelligent needs among departments, the provision of enterprises' intelligent systems is still aimed at a certain field of government affairs. For example, Ultrapower provides Chinese text analysis to the Local Taxation Bureau and converts it into institutional data. At the same time, it retrieves all kinds of data on the network regarding enterprise and shareholder relations, so as to facilitate the inspection of tax personnel.

Driven by AI technology, government services will develop in a more humanized and targeted direction. On the one hand, the public services for residents and enterprises will be more in line with people's habits, rather than relying solely on online interfaces. On the other hand, the AI decision-making will be more effective, accurate, and flexible.

Compared with the previous digital security systems, AI security presents two characteristics of real-time and intelligence, which improves public security management. The construction of intelligent cities in China has gradually reached a climax, and public safety as one of its core content has also gained broader development space.

Based on the role of AI in all links of public security, using AI technology for model training can improve police efficiency. At present, video surveillance is the main means of security. AI can participate in the visual information extraction process and build models, mainly regarding the characteristics and behaviors of target subjects, entourages, vehicles, and surrounding objects. It can obtain high-order semantic and strong expressive features and store them by categories.

According to security personnel's needs, AI can enact high-efficiency screening based on various conditions (e.g., vehicle, time period, area, scene pictures, phone numbers, ticket numbers, accommodation information) and quickly outline the target individual's action track. It is beneficial to improve case handling efficiency of police and realize the purpose of "using scientific and technological means to improve the police force."

Other security scenarios that AI can help to be effective in are: urban security (intelligent city), community security (intelligent community), campus security (intelligent campus), park security, plant security, etc. In addition, there are security systems designed for large-scale event sites such as concerts, airports, railway stations and other public transport hubs.

At present, with the help of AI technology, security has developed from passive monitoring to active early warning. AI and big data technology can be used to monitor large public places and roads. When the flow exceeds the threshold, it is reminded to take measures such as flow restriction to realize traffic flow control and management. Also, big data can be used to predict potential crimes and suspicious behaviors (such as purchasing contraband, squatting in a specific place, etc.) based on criminal records.

Digital government is a systematic project of slow work and meticulous work. It requires top-down decision-making in line with urban development and ensures maximum innovation around government data. There is no doubt that the construction of a digital government has been a key node with the prevalence of AI technology.

3.8 AI in the Legal System

The progress of AI technology is changing all aspects of human life in unprecedented breadth and depth. As an important force of the new generation technological revolution, AI will help to enhance the national and international status of a country.

Applying AI in the field of judicial trials is in accordance with China's national strategy. The recording and video recording of the whole judicial process will effectively realize intelligent monitoring of judicial power as well as reduce judicial arbitrariness, judicial corruption, and influence peddling.

In recent years, China's judicial reform has deepened, and the "smart court" represented by AI has been further clarified as the main embodiment of technical reform.

- **AI's Response to Judicial Requirements**

An AI judicial system is inseparable from the progress of technological conditions and the renewal of judicial theoretical frameworks.

On the one hand, digital technologies provide an essential precondition for AI integration in the field of judicial application. For example, after machine learning, the big data of each department will be classified and combined for case pushing. On the other hand, legal formalism provides theoretical support for the technological progress of AI. It follows the principle that the legal system is a closed logical self-sufficient conceptual system, and it bases its case analysis on laws and regulations before making a judgment.

Since 2013, especially since the establishment of the filing and registration system in China in 2015, the number of cases accepted by Local People's Courts at all levels has gradually increased. More complex and new cases are arising as well, such as virtual property disputes, data rights disputes, and information network security cases that rely entirely on information technology and can only occur on the Internet. These all need to be accepted by the court.

More cases means longer trial cycles. It reduces case trial efficiency and challenges the credibility of the court. Therefore, judicial system reform is in urgent need. AI technology has the functions of data analysis, classification, and retrieval. These functions enable fast processing of repetitive tasks and simple cases, and can greatly improve the work efficiency of judicial staff.

In addition, AI judicial reform is important in promoting the reform of a trial-centered litigation system. By strengthening data profundity and adding unified evidence standards to digital procedures, an AI judicial system is conducive to improving and maintaining social justice.

AI can internalize regulations and past representative trial cases in a very short time through Deep Learning. It can thus adjudicate based on screened evidence following legal rules and procedures.

Apparently, the promotion of an AI legal system can eliminate judicial corruption and injustice caused by human factors to the greatest extent. It will also optimize the judicial allocation and reduce unnecessary administrative staffing as well as financial burden. It will help create a fair and healthy judicial environment in all regions.

- **From Supporting Technology to Subversive Technology**

There are three main ways in which technology can reshape the judicial system. First, at the most basic level, technology can provide information, support, and advice to people involved in the judicial system. This refers to supporting technology. Second, technology can replace the functions and activities originally performed by human beings. This refers to alternative technology. Third, technology can change the ways judicial personnel work and provide different forms of justice. This refers to subversive technology, which is especially applied in areas involving significant procedural changes and predictive analysis can redefine the role of judges.

At present, most of the judicial reforms supported by technology focus on the first and second levels of technological innovation.

Innovations supporting technology in the first level allow people to seek judicial services and obtain information about judicial processes, choices, and alternatives (including legal alternatives) on the Internet. People are also increasingly looking for and obtaining legal support online. In recent years, the number of online law firms that provide "non- bundled" legal services has increased significantly.

On the second level, alternative technology such as digital video, video conferencing, teleconferencing, and e-mail can supplement, support, and replace many face-to-face live meetings. These technologies can support the judicial system and, in some cases, change the environment in which courts hold hearings. For example, online court procedures have been increasingly used in specific types of disputes and matters related to criminal justice.

Now, the combination of AI and justice has initiated changes and subversion in the third level. Under the background of database establishment, AI identifies the facts of cases through natural language processing and a knowledge atlas. It will then extract case information through a neural network, construct a model, and search for similar cases in a large number of databases for automatic push.

For example, the 206 system of Shanghai People's Court can form a sample of machine learning based on but not limited to: the subject of crime, criminal behavior, subjective factors of the offender, case facts, case dispute focus, and evidence. The sample will then be used to push cases for judicial personnel and provide a trial reference for judges.

The system integrates multiple data, analyzes the facts of the case from different angles, and makes legal judgments. In this way, intelligent machines are involved in and provide assistance to every process from case filing to court trial. In addition, the case trial auxiliary system can also learn to extract and verify the evidence information and predict the case judgment results by learning a large number of cases, so as to provide references for the judge's judgment.

In Mexico, AI is already able to make simpler administrative decisions. For example, the Mexican expert system provides advice to judges when "determining whether the plaintiff is eligible for a pension."

To sum up, the more important issue in judicial reform has changed from whether technology will reshape the judicial function to when and to what extent such changes will take place. Nowadays, AI technology is remodeling litigation affairs and the working mode of the courts. In the near future, more courts will continue to build and expand online platforms to support archiving, transferring, and other activities. These changes will provide a framework for the further growth of AI justice.

The concept of "Legal China" introduced by President Xi Jinping and the superimposition of Internet and AI technologies are providing greater impetus for the realization of the digital court. But in order to achieve this goal, a stronger foundation and further exploration will still be needed.

3.9 AI in Transportation

Only when technology is closely connected with human destiny can its revolutionary significance be fully displayed. The development of transportation has experienced several stages. At the very beginning, people travelled on foot. Then animals like horses, oxen, and donkeys were domesticated to be ridden as well as to pull carriages. Sedan chairs were also used at this time together with animal-driven vehicles. Later, with the emergence of the steam engine, cars and trains replaced these primitive means of transportation.

With the progress of AI technology, the value of an intelligent travel ecology related to the automobile is being redefined. The three elements of travel—"humans," "vehicles," and "roads"—are endowed with human-like decision-making capability and behavior, and this will cause great changes to the whole travel system. Powerful computing power and massive high-value data have become the core strength in building a multi-dimensional and collaborative travel ecology. As the application of AI technology in the field of transportation develops in the direction of intelligence, electrification, and sharing, the intelligent transportation industry chain centering driverless technology will gradually form.

- **The landing of driverless technology**

The first driverless car in human history, American Wonder, officially appeared in August 1925. It had a "flexible" steering wheel, clutch, brake, and other parts but with no driver on the driver's seat.

Behind this car, engineer Francis P. Houdina was sitting in a separate vehicle and controlling the American Wonder by transmitting radio waves. He "drove" it through heavy traffic in New York, all the way from Broadway to Fifth Avenue. However, this experiment, which can almost be regarded as "super large remote control," still hasn't been widely recognized by the industry today.

In 1939, skyscrapers began to spring up all over the US. People had gradually regained confidence after the Great Depression and were harboring a bright vision for the future. At the New York World Expo that year, everyone was rushing to see what the "future" looks like in Futurama (future world) built by General Motors.

Designer Norman Bel Geddes further explained in his book *Magic Motorways,* published in 1940, that humans should be freed from driving. American freeways should be equipped with something similar for motorcars to the train track that provides automatic driving guidance. Cars will follow a certain track and program when they are on the highways before returning to human driving after they are out of the high speed zones. The timetable he gave for accomplishing this idea was 1960.

But in the 1950s, when researchers began to experiment according to this assumption, they realized the difficulties were more severe than they expected. This was the time when a comprehensive exploration of realizing unmanned driving was carried out.

In 1966, intelligent navigation appeared in Stanford University. Shakey was the first robot with a wheel structure developed by the SRI AI research center. It had built-in sensors and software systems that create a precedent for automatic navigation.

In 1977, the Tsukuba Engineering Research Laboratory in Japan developed the first automatic driving vehicle that used a camera to detect front markers or navigation information. This indicates that people began to contemplate the prospect of unmanned vehicles from a "visual" perspective. Together, navigation and vision put the time led by "ground orbit school" to an end.

In 1989, Carnegie Mellon University pioneered the use of neural networks to guide self-driving cars, even though the server of the refurbished military ambulance in Pittsburgh was as large as a refrigerator and its computing power was only 1/10 of that of Apple Watch. But in principle, this technology comes down in a continuous line with today's unmanned vehicle control strategy.

China started research on intelligent mobile devices in the 1980s, and the initial projects also came from the military. In 1980, the state approved the project of "remote controlled anti-nuclear and chemical reconnaissance vehicles." Harbin University of Technology, Shenyang Institute of Automation, and the University of National Defense Technology participated in the research and manufacturing of the project. In the early 1990s, China developed her first driverless car.

Since 2000, automobiles have had more intelligent functions. GPS and sensors provided data and application support for unmanned driving. Technology and automobile manufacturers began to accumulate large-scale personal travel data, which enabled AI to learn driving essentials. Sensors allowed automobiles to have local real-time sensing and judging capabilities. For example, ABS (anti-lock braking system), airbags, and ESC (electronic stability control) all helped to improve the comfort and safety level of the vehicle. The real automotive intelligence began in the second decade of the 21st century. With Google taking the lead in AI technology, AI applied to automobiles were developed one after another. The main functions are reflected in various aspects including lane change, parking, and warehousing.

• The present and the future of unmanned driving

The Society of Automotive Engineering (SAE) categorizes unmanned driving in six levels.

L0—no automation: this is the stage in which the driver has absolute control.

L1—driver assistance: the system has at most "partial control" over choosing steering or braking. In case of emergency, the driver shall be ready to take over the control immediately at any time. Humans are also needed to monitor the surrounding environment.

L2—partial automation: different from L1, the control power transferred to the system in L2 changes from "partial" to "full." That is, in the ordinary driving environment, the driver can transfer both the horizontal and vertical control power to the system at the same time. Humans are still needed to monitor the surrounding environment in this stage.

L3—conditional automation: the system completes most of the driving operations. Only when emergency occurs does the driver give an appropriate response according to the situation. At this time, the system replaces humans and monitors the surrounding environment.

L4—high automation: the automatic driving system can complete all stages of driving operation without the driver's response. However, at this time, the system only supports some driving modes and cannot adapt to all scenarios.

L5—full automation: the main difference between this stage and the previous is that the system can support all driving modes. At this stage, the driver may no longer be allowed to be the subject of control.

At present, the intelligent driving technology at home and abroad is mostly at the level of L2 to L3. It is worth mentioning that, compared with L2 automatic driving, L3 automatic driving means that after the function is turned on, the vehicle will completely deal with all problems in the driving process, including acceleration and deceleration, overtaking, and even avoiding obstacles. It also means that in case of an accident, the responsibility identification will be officially changed from person to vehicle.

In other words, the automatic driving technology of L3 is an important dividing line between manned and unmanned driving, and it also indicates the transition between low-level driving assistance and high-level automatic driving.

L2 level automatic driving is mainly human-centered, with an auxiliary automatic driving system. L2 mostly corresponds to the common ADAS (Advanced Driver Assistance System) technology, including auxiliary driving functions such as ACC (Adaptive Cruise Control), AEB (Autonomous Emergency Braking), and LDWS (Lane Departure Warning System). The vehicle must be driven by the driver.

L3 is truly "unmanned." The autopilot system has completed most of the driving judgments and actions at this level. The vehicle engine system is in control under specific conditions, but the owner still makes decisions in case of emergency. The so-called conditions consist of several functional elements: expressway guidance (Highway Pilot (HWP), 0–130 km/h), traffic congestion guidance (Traffic Jam Pilot (TJP), 0–60 km/h), automatic parking, high-precision maps, and high-precision positioning.

The rise of unmanned driving is inseparable from the vigorous development of AI. During a driving experience, human drivers usually go through the following steps. First, observe the surrounding vehicles and traffic indicators and then enact a series of operations (e.g., accelerate, decelerate, steer, change lane, brake) according to their destination direction. This process is subdivided into a perception layer, a decision layer, and a control layer in the research of unmanned driving. According to the deduction, the combination of sensors, machines, and AI algorithms will completely surpass the process of human driving.

However, despite the existence of combined sensors that can conduct 360° full coverage scanning centered on the vehicle, this seemingly perfect deduction is in a technical dilemma. Mechanism intelligence represented by AlphaGo has proved that machines can far surpass humans in speed and accuracy. After the machine makes a decision, the signal can be quickly and accurately transmitted to the vehicle steering system, braking system, and transmission system. However, just as Professor Feifei Li from Stanford University, a top scholar in this field, emphasized when talking with historian Yuval Noah Harari, author of *A Brief History of Mankind* and *A Brief History of the Future*, that the existence of the world is far more complex than two groups. In addition to algorithms, there are many players and rules. As research in unmanned driving enters the deep waters, the problems of sensors, chips, and data are gradually being exposed.

Unmanned sensors collect information that supports and secures decision-making during driving. Although many sensors can surpass the human eye in a single index, the problem of fusion and the subsequent cost issues is holding back the progress of unmanned driving.

Another potential problem related to multi-sensors is chip performance. The more information we need about external road conditions, the more sensors we need to deploy. This demand puts forward higher requirements for fusion. In the case of high speeds, due to the change of road condition, the data will be more overwhelming.

According to Intel's calculation, a driverless car equipped with sensors such as a GPS, camera, radar, and laser radar will generate about 4 TB of sensor data to be processed every day. Such a huge amount must be supported by powerful computing equipment. Even the top GPU enterprises like NVIDIA are struggling with pushing the limit of computing power and power consumption. Therefore, in recent years, more special computing platforms have come into our view. There are the AI special chip TPU invested by Google, the BPU launched by Horizon, China's top start-up company, and Tesla's highly invested research on driverless chips. But it is safe to say that multi-sensor fusion will remain a huge technical obstacle for unmanned driving for a considerable period of time.

In addition to the technical bottleneck at this stage, unmanned driving will also need to deal with the conflict between business and technology in the subsequent commercial development. Theoretically, driverless technology can produce the greatest commercial value in the first and second tier cities. However, the current technology is inadequate for achieving this goal in one step, and more tests are needed to verify its safety and reliability. In short, the current road testing cannot promote the popularization of large-scale driverless vehicles.

Since unmanned vehicles on the road will face the trust crisis of users in case of accidents, a transition scheme has been made for self-driving cars to conduct closed site tests and open road tests in the suburbs (or new areas) of the city. At present, in order to ensure the safety of vehicles on the road, driverless vehicles must carry out simulation tests and closed site tests before gradually taking the open-road tests.

• Redefining Roads

Transportation is the lifeblood of economic and social development in the process of urbanization. Nowadays, great changes have taken place in our mode of transportation in terms of diversity of travel modes, convenience, comfort, and safety level. However, problems such as road congestion, difficult parking, and frequent traffic accidents still remain.

Traffic systems are time-varying, nonlinear, discontinuous, immeasurable, and uncontrollable. In the absence of data in the past, people studied urban road traffic in the state of "Utopia." But the development of instant messaging, IoT, big data, and other technologies have made full coverage of data acquisition and deconstructed transportation possible. A revolution in the transportation system has come.

The coordinated development of intelligent transportation has become a trend. The autonomous control ability of vehicles will be continuously improved, and fully automatic driving will finally be realized. It will change the relationship between people and vehicles, liberate people from driving, and provide preconditions for people to consume information in vehicles.

Vehicles will become an information node in the network that exchanges a large amount of data with the outside world. The network then changes the interaction mode between the vehicle, people, and the environment through real-time perception of the surrounding information and generating more types of information consumption.

In the future, with the popularization of automatic driving, most people no longer need to buy their own car. Transportation will become an on-demand service that assures full sharing of roads, cars, and other resources in order to improve the overall operation efficiency of society.

When both hands are liberated from the steering wheel, drivers are granted time and access to entertainment, information, the office, media and other applications. These new application scenarios will generate enough power to reshape the entire automobile industry and subvert existing concepts such as automobile ownership and liquidity.

Roads will be redefined. The future road will be an intelligent digital road. Every square meter will be encoded, and signals will be transmitted with active and passive RFID (radio frequency identification technology). The information contained in these signals can be read by the intelligent traffic control center and vehicles, and the underground road precision positioning of the parking lot.

Vehicle intelligence and automation will be the most basic requirements for the future road traffic system where human sensing ability is emphasized. Casualties caused by traffic accidents will be eliminated and the traffic-carrying capacity of the road network will be greatly improved. Of course, the basis for all this is to ensure that communication technology is high-speed, stable, and dependable.

By then, more advanced information technology, communication technology, control technology, sensing technology, and computing technology will be integrated and employed. The relationship between people, vehicles, and roads will be promoted to a new stage. The traffic in the new era will have the remarkable characteristics of real-time, accuracy, efficiency, safety, and energy saving. The intelligent transportation system will set off a technical revolution.

3.10 AI in Services

In recent years, with the help of AI interaction technology, robots have become significantly more intelligent and have gradually entered the stage of application landing. At this stage, given the application status of robots in special environments such as high altitudes, underwater, and natural disasters, the industry in China is divided into three categories: industrial robots, service robots, and special robots.

According to the preliminary definition given by International Federation of Robotics (IFR), intelligent service robot refers to an autonomous or semi-autonomous robot with service as the core principle. The difference between a service robot and an industrial robot lies in the different application domains. A service robot has a wider range of applications and can engage in various fields including transportation, cleaning, security, and monitoring except for industrial production.

Compared with industrial robots, service robots have a higher degree of intelligence. They mainly use intelligent control technologies such as optimization algorithms, neural networks, fuzzy control, and sensors to carry out autonomous navigation and path planning. Today, the value of service robots is made even more prominent with the prevalence of the coronavirus.

• From Substitute Assistance to Innovative Service

At present, service robots can be categorized as personal (family) service robots and professional service robots based on the differences of their application scenarios. Personal (family) service robots include domestic robots, leisure and entertainment robots, and robots for the elderly and the disabled. Professional robots include logistics robots, protection robots, field robots, commercial service robots, and medical robots.

From the perspective of product function, service robots can be categorized as replacing human beings, assisting human beings, and creating new fields. There are successful commercial models for each of these three categories. For example, Yunji China, a distribution robot to replace humans, industrial UAVs, and aerial drones to assist humans, and home escort robots in new fields.

In terms of replacements, the driving force of the industry derives from the automation demand of services, that is, "machine replacement." According to the National Bureau of Statistics, the proportion of China's tertiary industry in GDP has been continuing to increase, reaching 53.9% in 2019. At the same time, the tertiary industry absorbs more than 45% of the employed, which is 1.7 times that of the secondary industry.

In the era of industrialization, the automation demand of manufacturing industries such as automobiles, electronics and household appliances has driven the vigorous development of

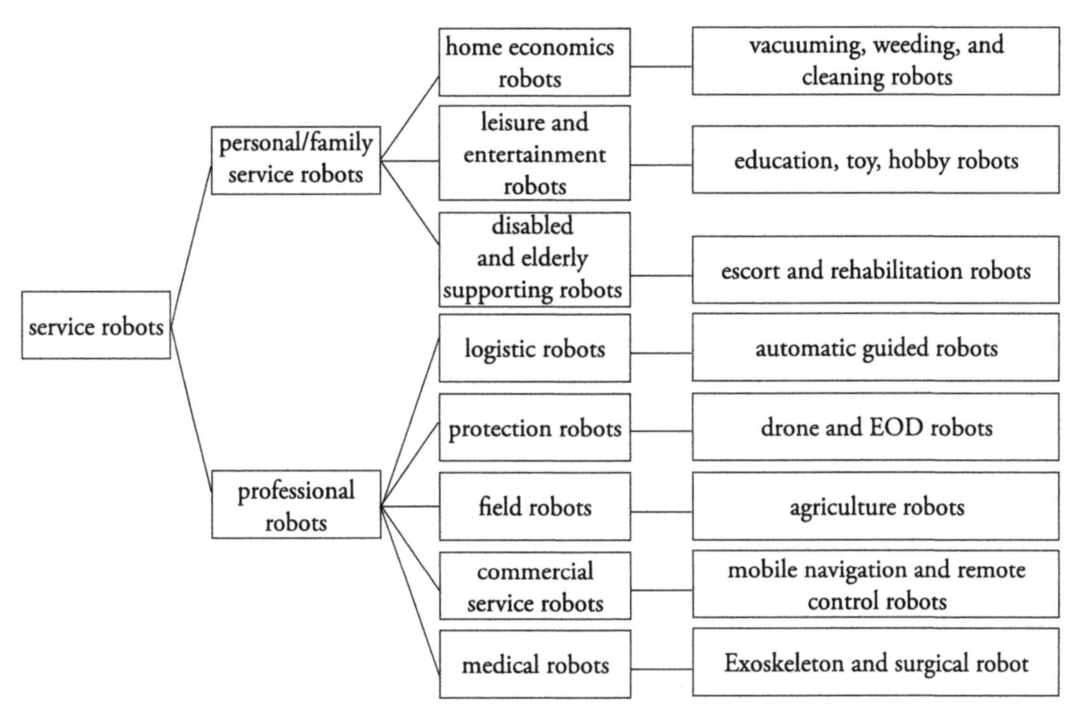

Source: IMRobotic, Huajin Securities Research Institute

industrial robots. In the future, with the rise of the tertiary industry, the automation demand in medical treatment, logistics, and catering is expected to accelerate the development of corresponding service robots. Particularly, in high-risk service industries, such as health care, rescue operations, fires, etc., the demand for machine replacement is stronger, as seen during the pandemic.

For example, the distribution robot not only reduces a medical staff's exposure to the infectious sources but also their labor intensity to a certain extent. In this epidemic situation, the distribution robots from Pudu Technology and Saite Intelligence have displayed desirable performances. They can deliver test sheets, drugs, and food according to the needs of the hospital, saving the energy of medical staff and reducing the risk of infection.

In addition, after the resumption of enterprises, the demand for intelligent distribution robots in office buildings will also increase significantly. Visitor flow rate in office buildings, and particularly in elevators, is high. So it is easy for people to cross-infect during meal times. Distribution robots or meal delivery robots can address this concern effectively. Some office buildings in Shanghai are already equipped with unmanned distribution robots. Couriers who pass the property temperature measurements can leave the takeout and express to the robot, which will then deliver the goods to each floor through gates and elevators by itself. This way, goods will not pile up at the delivery center, and buyers do not need to gather when collecting them.

In terms of assistantship, service robots can effectively improve people's work efficiency through AI, motion control, human-computer interaction, and other technologies. These robots do not replace humans but only exist as assistants. The price of such service robots is mainly based on their value to improve the human workload.

For example, with the acceleration of the rhythm of life, people hope to get rid of cumbersome housework. Therefore, domestic robots are invented to make life easier for us and to help us achieve our higher demand for quality of life. For younger generation consumers, smart products such as the smart phone, smart wear, smart homes, and smart cars are not only satisfy their living needs but also emotional needs. In a way, service robots are gradually becoming a part of people's daily life.

In terms of innovation, service robots begin to wade into "what people can't do" and 'what people don't want to do," so as to create new needs. For example, the application of professional robots in extreme environments and fine operations, such as Da Vinci Robot-assisted surgical system, anti-terrorism and anti-riot robots, military Unmanned Aerial Vehicles (UAV), etc.

Da Vinci Robot-assisted surgical systems can assist doctors in surgery, complete some extremely fine movements that cannot be completed by hands. The surgical incision can be made very small to speed up the postoperative recovery of patients. Anti-terrorism and anti-riot robots can be used to detect, remove, or destroy explosives in dangerous, harsh, and harmful environments. In addition, they can also be used in firefighting, hostage rescues, confrontations with terrorists, and other tasks. Military UAV can be used in various strategic and tactical tasks such as reconnaissance and early warning, tracking and positioning, special operations, precision guidance, information countermeasures, battlefield search and rescue, and other actions in the modern military.

• A Promising Future

With expanding application scenarios and modes, market demand for service robots is rising. According to the report of Chinese Institute of Electronics, the global robot market is expected to reach US$36.51 billion in 2021, with an average annual compound growth rate of 12% from 2013 to 2021. Among them, industrial robots take up US$18.14 billion, service robots US$13.14 billion, and special robots US$5.23 billion.

From 2013 to 2021, the average annual compound growth rate of global service robot sales was 19.2%, which was higher than the overall robot market. Meanwhile, the structural proportion of global service robots in the global robot market has increased year by year, and it is expected that the proportion will reach 36% in 2021.

For China, in 2019, the scale of the robot market reached US$8.68 billion, and the scale of the service robot market reached US$2.2 billion. It was a remarkable increase of its structural proportion. By 2021, it is expected that this proportion will reach 31.6%, and the scale of service robot market will reach US$3.86 billion, accounting for nearly 30% of the global market. It can be said that the service robot industry has ushered in a development boom all over the world.

At the same time, the development of 5G, AI, cloud computing, and other related technologies are promoting the evolution of the technology and function of the service robot. It is developing a more highly-sensitive perception, finer operation control, and more intelligent human-computer interaction.

5G technology is known for high speed, low delay, and large connections. It can transmit massive data in real time, provide network support for simultaneous applications, and promote robots to complete more complex and intelligent work. On the one hand, relying on 5G technology, relevant data can be quickly transmitted to the Cloud and provide data support for customer analysis (reception robots) and marketing decisions (new retail robots); on the other hand, 5G technology enhances the Cloud capability and provides a high-speed and stable network support for the function optimization of remote diagnosis, treatment of reception robots, and manipulator remote control.

In addition, the continuous development of Al technology with Simultaneous Localization and Mapping (SLAM), machine vision, voice interaction, and Deep Learning as the core can promote intelligent decision-making ability and scene coverage of service robots.

SLAM technology can accurately guide and locate service robots and realize autonomous obstacle avoidance. It will greatly improve the robot's work efficiency and delivery safety. With machine vision, functions such as visual active interaction and assisted navigation and obstacle avoidance in service robots can be realized; new retail robots can conduct intelligent demand analysis through face recognition. Finally, breakthroughs in natural language processing and Deep Learning will facilitate language and emotional communications, which will bring the development of reception robots and companion robots to the next level.

It is not hard to imagine that the future of service robots in the fields of intelligent management of public health systems, intelligent allocation of emergency materials, home companionship, and more is highly promising.

3.11 AI in Education

Human beings learn about the world around them with the help of tools, and the invention and innovation of tools drive human history forward. Similarly, the reform of education also means to promote the development of education. One of such innovations that are gaining popularity nowadays is AI intervention in education.

It is an inevitable fact that AI will change the concept of education, just as the birth and development of computer technology has rapidly and profoundly changed the way of human life in the fields of commerce, transportation, finance and production.

- **Integrating AI with Education**

The integration of AI and education is first reflected in the change of educational models and methods. The French scholar Monaco believes that the reform of education has been through four stages: the performance stage relying on direct transmission between people, the expression stage relying on indirect transmission of language and text, the image stage relying on sound and image recording, and the information technology stage relying on equal interactions.

At different stages, the methods of education are also different. This includes the method of learning (acquiring information), the method of teaching (disseminating information), and the method of teacher-student interacting. Apparently, technology is the key force driving educational reform both in the evolution of educational stages and in the changes of educational methods.

In the performance stage, the means of learning is relatively simple and depends completely on oral transmission. One typical form of education at this stage is the traditional Chinese private school, which is small in scale and has no personalized education.

In the expression stage and image stage, papermaking, printing, and imaging technologies begin to emerge. Teachers are no longer the only source of information, and teaching and learning start to gain relative independence from each other. At this time, schools at different levels are expanding in scale. Public schools address the average intellectual level of the majority population and advocate large-scale, efficient teaching. Knowledge dissemination is the main purpose of these schools built on top of the solid structure of teachers, teaching materials, and classrooms.

Modern information technology breaks through the space-time limitations of the same-position centralized education model. The empowerment of AI technology makes the transmission of knowledge faster and more equal. Profound changes have taken place in the teaching methods and modes, making the development speed of education faster than any period in modern history.

First of all, AI makes teaching and learning more targeted. It addresses individual needs of learners on a massive scale, and it completes the learning activity through learning management, learning evaluation, and cognitive thinking. AI can analyze students' learning methods and needs and then generate personalized learning programs that efficiently improve the learning experience and after-school tracking services. Teachers can improve the ways and content of their teaching based on collected feedback from students, and the intelligent evaluation system can also provide accurate intervention suggestions for teachers according to the situation of specific students.

Second, the AI education platform is the cornerstone of industrial intelligence. The construction of this platform requires student data collection and data in-depth analysis. In addition to completely tracking and recording students' online learning process, the platform will also record and store each student's actual data, including portfolio, academic performance, time data, knowledge, hobbies, reading data, etc. Then, the platform uses AI technology to predict students' learning preferences, specialties, intelligence level, struggles and demands, and finally extends to career development suggestions. All these practices will begin as soon as students enter school so that each student can be guided by a personalized education model.

The learning resources of the traditional education industry are often preset and can hardly meet the specific demand of every student. AI education, on the other hand, takes the personalized learning manual as the carrier to dynamically generate learning resources, which will further realize the goal of teaching students according to their aptitude. Improving the teaching closed loop will also benefit the distribution of educational resources and promote education equality and affordability.

Today, "AI + education" products and services are quickly put into use in preschools, K-12 education, higher education, and vocational education. The main application scenarios include photo taking and question searching, hierarchical course arrangement, oral evaluation, test generation and grading, composition and assignment commenting, etc. At present, "AI + education" is only an auxiliary link in the learning process. The more kernel learning links are, the longer it takes for them to be intellectualized. In the future, with the further development of educational surveying and AI technology, AI is expected to penetrate deeper into the core links of teaching and fundamentally improve learners' perceptions of and approaches to learning.

Finally, AI will change the school running form, expand learning space, improve school service levels, and form a more learner-centered learning environment. AI-enabled education governance will promote scientific education decision-making and precise resource allocation, and accelerate the formation of a modern education public service system.

• Education Driven by Technology Is Not Equal to Technology

The prevalence of AI intervention in education indicates that the ongoing education reform calls for the support of new technologies. However, the history of technological development shows that new challenges always arise during this process. These problems are often not limited to the technical aspects but more relevant to the humanistic and ethical realms.

Education is a systematic process that cultivates individual autonomy, etiquette, and responsibility. Education is also an inherited structure of intelligence and knowledge through which collective identification is established.

Information technology can contribute to education efficiency. But it is important that these technologies ensure good learning outcomes that prioritize human dignity rather than costs and arithmetic measurement. In the era of AI intervention in education, we must reconsider the definition of education—what kind of education is a good education?

Education technicalization does not mean "pan technicalization," and the technical rationality of education is by no means instrumental education. In the integrated development of education and technology, technology not only offers innovative means for educational modernization but also makes educational objectives easier to achieve. While people generally feel positive about new technologies and actively engage in their application, over reliance on "technical support" and blind pursuit of "technological innovation" often lead to arrogation of technical rationality to value rationality.

News about using AI to monitor students' learning status or implement mandatory control has triggered heated discussions. Blind obsession with learning control through technical methods overlooks the questions such as "why to control" and "what kind of management is good management" that represent a source of value and ethical norms of educational management.

Although AI technology can manage learners' behavior and cognition in a detailed and comprehensive manner (in fact, all learning factors can be quantitatively assessed by powerful algorithms), we must know that technology is only the reflection of an ideal way of living that includes the most important but often ignored questions.

In other words, under the intervention of AI in education, people regard learning results as profit margins, computing technology, and infrastructure as dividends of return on investment, and even take students as industrial resources to evaluate their ability to acquire "work skills." By doing this, we gradually underrate the ability to create meaning through reflection and contextualization. Instead of emotion and sensibility, people tend to pay more attention to the accuracy of rationality.

Education carries hope. People always hope that their input can lead more quickly to a splendid output. The utilitarianism of education, meaning that people place more emphasis on the external goal of education over its internal value, has encouraged quick delivery of "educational products" and turned education into competition.

In the era of AI education, when education is shaped by "technological innovation" and "new reform," learners are also required to constantly adjust their learning to adapt to the changes of society and educational activities. In this process, the purpose of education is not simply to enrich "human nature" but also to adapt to the market.

There is no doubt that technology is significantly beneficial to social development. Without the support of technology, there cannot be a human-oriented world. "Human civilization" is the embodiment of the internal value of technological progress. However, in the era of a close connection between technology and social life, we should be more vigilant against the social changes brought by technology. Human education should not be constrained by technology, nor should it be defined by technology.

Education is technology driven, but education does not equal technology. In view of this, education should not only actively take technological progress as an effective driving force of reform but should also maintain the bottom line of its value origin during the integration process. Moral teaching with the purpose of making learners better human beings should always be the foundation of any kind of education.

3.12 AI in Creation

AI technology application has landed in industries of all walks of life. In addition to the field of manufacturing, it also triggers an unprecedented revolution in the field of aesthetic art, which has been regarded as an exclusive reflection of "human originality" in the past.

Nowadays, AI no longer represents simple imitation of human intelligence. In particular, AI with computational intelligence, perceptual intelligence, and cognitive intelligence are not only qualified for enacting automatic driving and visual recognition but can also operate natural language processing through Deep Learning and participate in innovative production as a creator.

However, this prospect has also aroused widespread controversy. Unlike mechanized production, AI creativity poses direct challenges against the unique status and value of human beings, and it further leads to human existential anxiety and the question of whether AI will one day replace humans entirely.

There is no doubt that in the future, everyone will be surrounded by a variety of highly terminal AI products. The best way to foresee the future is to create it. In the face of a world with new possibilities for us to explore, it is crucial that we understand the meaning of each new boundary we make.

• AI Reconstructs the Rule of Creation

For a long time, scientists have been striving to make the computer process human language, in an attempt to make intelligent creation possible such as conducting in-depth analysis of lexicon, sentences, and paragraphs.

In 1962, the earliest poetry writing software "auto-beatnik" was born in the United States. In 1998, "novelist Brutus" was already able to produce a short story with a reasonable plot connection in 15 seconds.

In the 21st century, the collaborative creation of machines and humans became more common. Different writing software was invented one after another, and users can have their works automatically generated for them simply by typing in keywords. Typical examples of this type of software are the "Jiuge computer poetry creation system" of Tsinghua University and the "Microsoft Couplet" developed by the Microsoft Asia Research Institute.

In 2016, Japanese researchers presented short stories at the Hoshi Shinichi Literature Competition, where they stood out among all entries and were successfully shortlisted. This suggests the possibility of AI composition achieving the same literary complexity and profundity as human writings.

In May 2017, "Microsoft Xiaoice" published the first AI poetry collection "Sunshine Lost the Glass Window." Some poems were published in *Youth Literature* and other publications or on the Internet. Xiaoice announced that it enjoyed the copyright and intellectual property rights of the works. In 2019, Xiaoice and human poets jointly created a poetry collection titled "Flowers are the Silence of Green Water," which is also the world's first literary work jointly created by intelligent machines and humans.

June 29, 2020, was a special date for Xiaoice. On this day, the AI software was awarded the title of "honorary graduate" of the year 2020 along with its human classmates at the Music Engineering department of Shanghai Conservatory of Music.

The example of Xiaoice indicates that the world has entered a new era of vigorous deepening of human-machine collaboration and aims for the continuous improvement of the quality of the work. AI creative practices also objectively promote the changes of existing art production modes and provide the necessary technical groundwork for the new art forms.

On the one hand, as a new technical tool and media of artistic creation, AI has renewed the concept of artistic creation and injected new development vitality into contemporary art practice. The impersonal intelligent machine "Quick Pen Xiaoxin" can complete news reports that generally take 15–30 minutes for human writers within 3–5 seconds. Similarly, "Jiuge" can generate seven character rhythm poems, acrostics, or five character quatrains in a few seconds. It is obvious that unlimited storage space, inexhaustible creative enthusiasm, and tireless learning ability with the unlimited expansion of corpus are absolute advantages of AI over human brains.

On the other hand, AI is reconstructing the boundaries of creative subjects in the process of collaborative text generation with human authors, and will become the leader of machine authors with higher degree of personalization in the future. For example, developers of Microsoft Xiaoice claim that it is not only capable of recognizing pictures and sounds but also has strong creativity and EQ based on in-depth learning. These features make Xiaoice essentially different from the intermediate form machines in the previous decades. As Xiaoice explains in its poem, "In this world, it is meaningful that I am beautiful."

• AI challenges human creativity

The most important controversy surrounding AI creation lies in the challenge it poses against human nature.

In traditional creative activities, humans as the subjects of creation are often deemed the authority and owner of inspiration. In fact, it is precisely human's radical creativity, irrational originality, and even illogical laziness rather than stubborn logic that make it difficult for machines to imitate human creation even now. Creativity is still an exclusive right of human beings, and there has never been a precedent of non-human creative subjects in reality.

However, with the emergence and development of AI creative production, this understanding of human beings as the only subjects of creation is gradually undermined.

From the perspective of imitation, AI can create works with similar style through the imitation of existing works. In this case, artistic creation still has its unique existence value, but human existence is no longer needed in the stage of repeated creation or mechanical reproduction. It is like the emergence of machines in the industrial age, which replaced labor and further improved productivity.

But it is undeniable that even the possibility of AI imitating the form and style of artistic works makes the role of creator no longer a human patent. The replicability of "creation" continues to weaken the influence of the original author or creator in the product. Language, being the omnipresent participant in the creation process, provides the medium for AI to realize its imitation, making the complete exclusion of human participation possible.

But at present, AI creative behaviors are not conscious. Further concerns are expressed about the change from consciousness flow to data flow or the functional substitution of data flow for consciousness flow.

Right now, AI is still in the process of learning for supplementing human intelligence. There is a trend of conducting professional research in the field. When AI does achieve the capability of replacing humans in terms of professional skills and apply it in interdisciplinary scenarios, it will undoubtedly become "human like" or even "beyond human." This means that AI development will soon shift from focusing on technology to subjects.

In other words, AI is growing stronger. First, highly anthropomorphic interactions are penetrating into all aspects of people's social life. AI is being personalized through the construction of comprehending human emotions. Like Siri, different types of accompanying robots are developing in-depth adapting systems in terms of emotions and human nature.

Second, the main body of AI does not depend on the technology in one specific field. Rather, it is the integration of natural language processing, computer vision, voice processing, and other technologies.

Third, the number of AIs is growing exponentially. Alexa, for example, receives the most hardware coverage from Amazon, and Microsoft Xiaoice has the largest AI interaction volume in the world.

Back to the discussion of AI challenges, we will find that the entry of AI into the creative field is not a negation of everything with "human" as the core. AI technology brings people a new way of thinking about artistic creation, but people remain unchanged. Therefore, creators should pay more attention to the development of domains that center creativity, ideology, and distinctiveness.

The progress of technology will bring subversive changes to the world of art. It is worth thinking about how aesthetic art can gain new life in the present technological revolution. In the future, human and intelligent technology will achieve a higher degree of integration, and the application effect of AI technology in art will overturn the current human perceptions and imagination. How to reconstruct the laws of beauty and combine creation and interaction is worthy of thorough contemplation.

CHAPTER 4

Trends and the Future

4.1 The Time of Computing Power

Times are changing. Data, algorithms, and computing power have become the key words in the era of AI.

Internet popularization promotes digital device connection, and the development of IoT further facilitates this process by creating data accumulated in geometric series with massive equipment and superimposed complex application scenarios. The International Data Corporation (IDC) pointed out in *Digital Era 2025* that global annual data production will increase from 33 zettabytes (ZB = 10^{21} bytes) in 2018 to 175ZB in 2025. In other words, the age of big data is arriving.

Explosive data nurtures AI development. Algorithms that have been difficult to practice in the past, like Deep Learning, are now able to be trained and implemented on an extensive scale. Therefore, higher end computing power is needed.

In a way, computing power reflects civilization's advancement. Only gradually did the means of calculation, which took the form of knotting in primitive societies and the abacus in agricultural societies, evolved into the computer in industrial society.

The development of computer calculations has also experienced several stages, from relay computers in the 1920s to electron tube computers in the 1940s, and then to diode, triode, and transistor computers in the 1960s. Among them, transistor computers can calculate at a speed of hundreds of thousands of times per second. With the emergence of integrated circuits in the 1980s, the computing speed reached tens of millions of times per second. Now, it has further multiplied to billions, tens of billions, and hundreds of billions of times per second.

Research on human biology shows that there are six layers of the cerebral cortex, and the neural connections in them form a geometric progression. The neural synapses of the human brain beat 200 times per second, while the cerebral nerves beat 14 trillion times per second, making this number the inflection point for computers and AI to surpass the human brain. In this light, the progress of human wisdom is related to the speed of computing tools that people invent. Computation power is the core of human wisdom.

In the past, computing power was mainly regarded as a kind of calculation capacity. But the era of AI has enriched its connotation, including the technical ability of big data, the ability to provide problem-solving instructions, and the ability of systems computing programs. Basically, computing power now refers to the ability of computer programing. It is a limited, determined, and effective method to solve problems with computer programs—the basis of computer science.

Computer power consists of four parts. The first is the system platform, which is used to store and calculate big data. The second is the central system, which is used to coordinate data and business systems. The third is the scenario, which is used to coordinate the application of cross-departmental cooperation. The fourth is the data cockpit, which directly reflects the data management ability and application ability. When it is applied in practical problem solving, computing power changes the existing mode of production and helps whatever/whoever is using the computer in decision-making and information screening.

With diversified scenario applications and iterative new computing technologies, calculation activities and computing power are no longer limited to the data center and can also happen in the Cloud, in networks, edge, and end scenarios. Computing begins to grow independent of its tools and physical properties and is evolving into a ubiquitous capability.

From the perspective of function, with the continuous upgrading of human demands for calculation, computing has gradually become capable of perceiving, natural language processing, reflecting, and discerning on the basis of its single physical tool attribute. With the help of a series of digital software and hardware infrastructures, such as big data, AI, satellite networks, optical fiber networks, the IoT, cloud platforms, and near-earth communication, computing technologies and products are penetrating into all aspects of social production and life.

Whether the small electronic products such as laptops, smart phones, and pads or expanding applications in civil fields such as weather forecasting, border travel, medical security, and clean energy, they are inseparable from the enabling support of computing. Computing has now achieved a complete transformation and become an extension of human capabilities in enabling the digital transformation and upgrading of all walks of life in the digital economy.

As Nicholas Negroponte says in the preface of *Being Digital*: "Computing is no longer just about computers, it also determines our own existence." Computing power is becoming an increasingly important factor in people's social life styles.

Computing power can make AI smarter and more industrialized. In the future, as AI becomes more powerful and more fundamental, its requirements for computing power will also grow accordingly.

In April 2020, the National Development and Reform Commission defined the scope of "new infrastructure" for the first time. It includes computing infrastructure represented by data and intelligent computing centers. According to the report, the AI computing power reflects China's cutting-edge innovation ability, and the investment in this field also shows the country's attention to AI at the strategic level as well as the enterprises' urgent wish to improve their core competitiveness.

4.2 Ubiquitous Intelligence in Social Life

In 2020, the year of the global anti-epidemic battle, AI quickly landed from the Cloud and played a significant role in medical treatment, urban governance, industry, non-contact services, etc. AI has never been more closely connected with industries. Once again, the value of AI as an important social driving force in the new digital revolution and industrial change has been verified.

On July 10, 2020, at the Tencent forum of the World AI Conference, Si Xiao, the Vice-president of Tencent and President of the Tencent Research Institute, officially released the *Tencent AI White Paper*. It outlined the panorama of ubiquitous intelligence from the five dimensions of macro background, technical research, landing applications, future economy, and systems security.

Since the Dartmouth Workshop in 1956, AI development has experienced several ups and downs. Nowadays, although "strong AI" with real perception and self-consciousness is still a fantasy, "weak AI" focusing on specific functions has been highly achieved and has sprung up like mushrooms.

In terms of technology alone, the tendency of AI being machine learning and Deep Learning-dominated has been considered one of the most important technological and social changes faced by mankind. Machine training has become the new starting point of the second generation technological social form since the birth of the Internet. In the past decade, the computing resources for AI training models have increased sharply. Each year between 2010 and 2020, the computational complexity of AI has multiplied by 10 times while the training cost has decreased by about 10 times.

In regard to computing power, it is greatly improved thanks to the improvement of microcircuits processing capacities and the decline of hardware prices. Multiple AI technologies have found clear application scenarios based on these preconditions. According to the data provided by Tsinghua University, the market scale of computer vision, voice technology, and natural language processing accounted for 34.9%, 24.8% and 21% respectively, which are the three largest application directions in China's market.

From the perspective of applications, given the rapid development of computer vision, image recognition, natural language processing, and other technologies, AI technologies have been widely applied in many vertical fields and scenes to provide diverse products and solutions. In recent years, the world's largest technology companies, such as Apple, Google, Microsoft, Amazon, and Facebook have all invested more and more resources to seize the AI market and even transformed into AI driven companies completely.

The COVID-19 pandemic has been a good litmus test of AI capacity. AI companies began to actively play a role in improving the overall efficiency in confronting the virus. AI applications in medication span across image recognition to medical screening to help new drug research and development. In addition, AI technology promotes remote consultation and online popular education of medical information so that people can access medical resources more quickly.

When the pandemic ends, there will be no exclusive "traditional industries." Each industry has more or less started the digitalization process. Due to the impact of difficult employment, increased costs, and laborers infected with COVID-19, manufacturing and service industries are accelerating the process of man-machine integration and being more intelligent.

In the epidemic prevention and control, AI technology is widely applied in urban governance, which also shows that China's intelligent society is gradually emerging. Basically, the pandemic has offered rich scenarios for AI practices to bring the vision of ubiquitous intelligence to life more quickly.

On the one hand, ubiquitous intelligence can be found in infrastructure construction, which is currently one of the seven major focuses of "new infrastructure construction" project and the general basic technology in the process of informatization in China. AI technologies will gradually transform into fundamental services like the network and electricity and provide common AI abilities for all industries. In the era of the Industrial Internet, it is the key for industrial digitalization, upgrading, and revolution.

On the other hand, ubiquitous intelligence can be applied in more multi-modal scenarios with larger receiver populations. With the continuing optimization of technologies, algorithms, scenarios, and human resources, the value of AI technologies in various fields (e.g., industrial, medications, metropolitan) is further verified. Undoubtedly, more innovative, AI-integrated industries will be invented in the near future.

4.3 Internet-AI Integration and Co-Development

In most cases, the third AI upsurge is attributed to rich big data resources, AI algorithm innovations, and great improvement in computing power. What is often ignored, however, is the important contribution of the Internet and Internet enterprises.

It is no accident that the world's leading Internet companies, including Google, Amazon, Facebook, and Alibaba, are also leaders in the field of AI. In fact, Internet enterprises are not only a booster but also an important protector for AI development.

First, Internet companies are innovators and practitioners of digital economy. These companies create and accumulate large amounts of data in production and business activities that come from users' real needs, feedback, and behaviors. On the basis of security and compliance, Internet companies not only made full use of data value but also activated the data awareness in the entire business realm and brought data value to a higher level. They promoted the penetration and development of the digital economy and completed the accumulation of big data resources during the third AI spring.

Second, Internet companies are in urgent need of AI technologies to realize the ultimate demand of economic digitalization. An enterprise that provides services online depending on data and algorithms must be empowered by AI technologies, leading to a spillover effect in the industry as they are introduced to more fields.

Third, the market expands greatly as the acceptance of AI technology in the whole society increases. For example, the cumulative sales of Tmall Genie, the smart speaker that enables intelligent entertainment, knowledge acquisition, and interactive experience for family users (now business users as well) had exceeded 10 million units by January 11, 2019, meaning that 10 million individuals or families have experienced the power of AI. In addition, Apple Siri, Microsoft voice assistant Cortana, Amazon's Echo smart speaker and Alexa voice assistant, Alipay's brush face payment also play an active role in the expansion of the AI application market.

Finally, the world is now experiencing an evolutionary moment "from Internet to AI." The rise of AI is largely the end product of the intersection of algorithm technology innovations and the Internet platform. Under the bearing of the large-scale public Cloud, the combination of the Internet, big data, and AI will extend to the physical world through the IoT and set the general tone for the future development of AI industries.

Based on data accumulation, technology spillover, and algorithm innovations, the Internet enables the connection between AI innovators and consumers; and, with the support of the public Cloud, which carries AI technology spillover and empowerment, completes the closed loop of the two-way feedback between data and intelligence that confirms the actual landing of the third AI wave.

4.4 Create a New Engine for Economic Development

AI's influence on business ecology is mainly reflected in three levels.

The first is reform in enterprises. AI technologies will participate in enterprise management and the production process. The trend of enterprise digitalization is becoming more and more obvious, and some enterprises have realized more mature intelligent applications. These enterprises have been able to collect and utilize multi-dimensional user information through various technical means and provide targeted products and services to consumers. At the same time, by optimizing the data, they gained insight into developmental trends so that they could better meet the potential needs of consumers.

The second is reform in industry. The changes brought about by AI technology have fundamentally changed the upstream and downstream relationship of the traditional industrial chain. The types of upstream product providers will increase, and users will also change from individual consumers to enterprise consumers or combined consumers.

The third is reform in human resources. New technologies will improve information utilization efficiency and reduce the number of employees. In addition, the wide application of robots will replace the labor force engaged in process work, resulting in an increase in the proportion of technical and management personnel and changes in the human structure of enterprises.

As AI industries represented by smart homes, smart networked vehicles, and intelligent robots become important engines to drive economic growth, Price Waterhouse Coopers proposed that by 2030, AI will increase the global GDP by 14% and contribute 15.7 trillion dollars to the world economy.

AI reconstructs the mode of production organization and optimizes the industrial structure on the basis of digitization and networking. It also opens up new space for economic growth by promoting multi-industry innovation and productivity. Accenture predicts that by 2035, China's labor productivity will increase by 27% and the total economic added value by 7.1 trillion dollars.

The value of AI landing industry is manifested in three aspects: automation, intellectualization, and innovation. Automation means relying on AI technologies to improve the degree of business automation. Machines will replace humans in the original business process to improve efficiency and reduce costs.

For example, industrial robots can take over repetitive labor actions such as sorting and assembly. They can help with film reading in medical imaging to improve the diagnostic efficiency. They can also be used in procedural advertising in the field of advertising marketing, etc. In most scenarios, automation refers to the activity in one single link in the business chain.

Intellectualization is to enable machines the ability of analysis and decision-making based on cognitive intelligence technologies like knowledge maps, so that they can accomplish tasks beyond the reach of human labor.

For example, to look for hidden relationships among hundreds of millions of entities and find gang crimes based on the industry knowledge map technology; or to achieve extremely high accuracy in sales volume prediction and reduce inventory and loss based on machine learning. Intellectualization mainly addresses the work of analysis, reasoning, and decision-making. The applications that are usually involved are cognitive intelligence technologies such as data mining, NLP, and Deep Learning.

Innovation is to reshape business processes and industrial chains after the deep integration of AI and industry, and form new business models and even new industry segments.

For example, the cost of an intelligent container based on computer vision is reduced by more than 50% compared with traditional mechanical vending machines, and it can accommodate more kinds of goods. At present, unmanned vehicles are the landing direction of AI with the most innovative potential. Once the driverless technology is mature, the industrial chain relationship between the main engine plant and the vehicle scene in the traditional automobile industry will be subverted.

4.5 AI is Understanding Humans

For a long time, the possession of emotions has been one of the important criteria to distinguish between man and machine. In other words, whether the machine has emotion is a key factor for assessing its degree of "humanization."

At present, AI technology is expanding both vertically (e.g., math operations, logical reasoning, expert systems, DL) and horizontally (e.g., smart phones, smart homes, intelligent transportation, and intelligent cities).

As "perceptual intelligence" gradually changes to "cognitive intelligence" with the ability of understanding and expressing, the next step will be to endow emotion to machines. Marvin Lee Minsky, the father of AI, once said, "If machines cannot simulate emotions well, people may never think machines have intelligence."

The attempt to make AI understand human emotions is not recent.

As early as 1997, Professor Rosalind W. Picard of the Media Laboratory of MIT put forward the concept of emotional computing. She pointed out that emotional computing is derived from emotions or computing that can affect emotions. In short, emotional computing aims to enable computers with the ability to recognize, understand, and express people's emotions, so as to make computers more intelligent. After that, emotional computing began to appeal to researchers in information science and psychology as a new area of study.

Emotional computing involves emotional recognition, emotional expression, and emotional decision making. Emotional recognition means letting machines accurately recognize human emotions with no uncertainty and ambiguity. Emotional expression means letting machines express emotions in the most appropriate form, including language, voice, posture, and expression. Emotional decision making means a study of how to make better decisions by using an emotional mechanism.

Emotional recognition and expression are the keys in emotional computing. Emotional recognition obtains the emotional feature data that can represent human emotions to the greatest extent by extracting the feature of an emotion's signal. A model will then be built to show the mapping relationship between the external representation data of emotions and the internal emotional state. This is how people's current emotional types are recognized. The methods include speech emotional recognition, facial expression recognition, and physiological signal emotional recognition.

Facial expression recognition is undoubtedly a crucial component of emotional recognition since 55% of human emotions are transmitted through facial expressions during communications.

In the 1970s, American psychologists Paul Ekman and Wallace V. Friesen pioneered in the study of modern facial expression recognition. Ekman defined six basic expressions of humans: happiness, anger, surprise, fear, disgust, and sadness. He also determined the category of recognition objects and established the Facial Action Coding System (FACS), which allowed researchers to describe facial actions according to a series of facial action units and detect subtle facial expressions according to the relationship between facial motions and expressions.

So far, emotional recognition is the most promising application of AI. For example, commercial companies can use emotional recognition algorithms to observe consumers' expressions when watching advertisements. They can thus better predict the sales prospects of a certain product, whether it will rise, fall, or remain the same, and prepare for its next-step development.

In addition to recognizing and understanding human emotions, machines also need to give emotional feedback, that is, the emotional synthesis and expression of machines. Similar to the ways humans express emotions, machine emotion expressions can be achieved through multimodal means such as voice, facial expression, and gestures. Therefore, machine emotional synthesis can be divided into emotional speech synthesis, facial expression synthesis, and body language synthesis.

Among them, speech is the most powerful and accessible means of emotional expression. The emotion of speech mainly relates to its content and the characteristics of the voice itself. Obviously, users will feel better if the voices of machines are with feelings.

If we look at the decision-making process from the perspective of emotional computing, we will see that people do not always depend on reason when solving problems. Therefore, inputting emotional variables in the process of AI decision-making can also help the machine make more humanized decisions.

Microsoft researchers have proposed a new reinforcement learning method based on the intrinsic reward of Peripheral Pulse Measurements, which is related to the responses of the neural system. Researchers assume that this reward function can help with sparse and skewed reinforcement learning and improve sampling efficiency.

Emotional intelligence not only makes machines more powerful and effective but also closer to human values. On May 29, 2014, the basic framework of AI interactive agent Xiaoice, developed by Microsoft's Asian Internet Engineering Institute, was publicly beta-tested on Wechat and won the favor of more than 1.5 million WeChat groups and more than 10 million users in three days. We can say that Microsoft Xiaoice is an AI capable of emotional computing at a preliminary stage.

Li Di, the head of the Xiaoice development team, once said that the AI platform has built a positive interactive cycle among technology, products, and data. In other words, Xiaoice has accumulated enough big data to accomplish self-evolvement.

Today, with 10 billion rounds of dialogue information with humans and a large amount of historical data, Xiaoice can anticipate the content of future dialogue with an accuracy above 50%. In a way, Xiaoice has formed a preliminary ability of memorizing, cognition, and consciousness.

It can be anticipated that future emotional computing will subvert the traditional human-computer interaction mode and realize the emotional interactions between humans and machines. The paradigm shift from perceptual intelligence to cognitive intelligence and from data science to knowledge science will also be better illustrated with AI development.

CHAPTER 5

Risks and Challenges

5.1 Algorithm Black Box and Data Justice

With the personalized push of TouTiao's headlines, the "sesame" credit score of Ant Financial, the "daddy-in-charge index" of JD, and "frequent-visitor rip off" phenomena with big data on tourism websites, we see more and more AI applications based on personal data and machine learning aiming for Cloud computing—in other words, automatic analysis of personal information—is being normalized in our daily life.

As more and more data are generated, the role of the algorithm has gradually changed from a single mathematical analysis tool in the past to a force that can leave an important impact on society. The algorithm based on big data and in-depth machine learning will be increasingly powerful in carrying out autonomous learning and decision-making.

With stronger features in generating new knowledge and rules through existing knowledge, algorithms are influencing the markets, society, government, and individual lives on a large scale. On the one hand, it makes services such as investment consulting and medical treatment easier and more accessible; yet on the other hand, it can make mistakes and even be "malicious" due to its dependence on big data, which is not neutral

• Data Is Not Unjust

Generally speaking, an algorithm is an operation program that analyzes and calculates certain data to solve specific problems. At first, an algorithm is only used to analyze simple problems on a small scale. An algorithm needs to have the following basic characteristics: input-output, universality, feasibility, certainty, and finiteness.

The essence of an algorithm is to obtain, occupy, and process data information and generate new data information on its basis. In short, the algorithm is the transformation and reproduction of data information or all knowledge obtained.

The algorithm follows the technical logic of "deriving" definite and repeatable facts and rules from structured facts and rules. For a long time, people believed that this algorithm technology born out of big data technology was ethically neutral.

However, with the third AI upsurge, the acceleration of its industrial and social application and innovation, and the growth of data levels are overturning this assumption. People realize that big data, which is extracted from human society, is inevitably inheriting the features of inequality, exclusion, and discrimination.

Deep Learning triggered the wave of the third AI fever. At present, it is used in the most excellent applications like AlphaGo. Different from traditional machine learning, DL does not follow the process of data input, feature extraction, feature selection, logical reasoning, and prediction. Instead, the computer automatically learns and generates advanced cognitive results directly from the original features of things.

But between the input data and the output answers of DL, there is an unknown "hidden layer" called a "black box," inaccessible to humans both visually and intellectually. It means that people cannot comprehend even if the computer tries to explain to us.

As early as 1954, Jacques Ellul pointed out in *The Technical Society* that it is superficial and impractical for people to believe they can control technology simply because they invented it. How technology actually develops in real life is usually incontrollable, even by technicians and scientists.

Ellul's words are preliminarily verified now in the era of AI, where algorithms are rapidly growing and evolving by themselves. DL also highlights a technical barrier brought by the "algorithm black box" phenomenon. It has become difficult to identify the cause of a certain breakdown whether it is because of program error or algorithm discrimination.

• Price discrimination and Algorithm Bias

Due to an algorithm's fundamental demand for data, a wealth of information elements with profound economic and social importance have been derived. The mastery and analysis of personal information is now simple and commonplace, making people the object of calculation. The derived algorithm discrimination includes price discrimination and algorithm bias.

The economic definition for price discrimination is the price difference between two or among more identical goods with the same marginal cost of production. This price difference lacks a cost basis.

There are certain preconditions that need to be met for an enterprise to successfully implement price discrimination. First, operators have considerable market power. Second, operators have the ability to predict or identify consumers' purchase intentions and potential. Third, the possibility of resale arbitrage is excluded to prevent consumers who enjoy low prices from reselling and undermining the effect of price discrimination.

Primary price discrimination means that the seller determines the upper limit of the buyer's willingness to pay as the goods' selling price. In this case, the seller can maximize the profit when charging each buyer. The more comprehensive operators grasp consumer information, the greater their ability to implement price discrimination becomes and the higher profits they obtain.

Secondary price discrimination refers to the provision of different versions of the same goods or services. The seller often knows little about the characteristics of the buyer and provides a series of sales agreements including price and various terms for the buyer to choose.

Tertiary price discrimination means when the seller determines prices according to the demand elasticity of different buyer groups. The tertiary price discrimination is more frequently seen in reality. Different prices charged by cinemas or scenic spots for students, the elderly, or minors are two examples.

Prior to the era of industrial intelligence, studies of the economy paid more attention to the secondary and tertiary price discrimination. Because it was difficult for the seller to accurately grasp the reserved price of each consumer, primary price discrimination didn't often occur. However, with the expansion of data scale and optimization of algorithm analysis, primary price discrimination is now possible to achieve in practice as a popular business strategy.

Obviously, in the era of big data, sellers, if their algorithms are powerful enough to adequately collect comprehensive information to discern the possible intention and capability of each individual consumer, can separately set different prices for consumers. With the help of big data technology, sellers can detect which users are capable of paying higher prices and which users are not, thus giving rise to the "big data rip-off."

There are three common "ripping off" routines. Differential pricing for different devices (e.g., Apple users vs. Android users); different pricing for different locations (e.g., the price set for users far from the mall is higher than for those that are closer); difference pricing for different consumption frequency—generally, the higher the consumption frequency, the higher the users' price tolerance.

"Big data rip-off" occurred as early as 2000. An Amazon user found that the price of a DVD he had previously browsed changed from $26.24 to $22.74 after deleting browser cookies. Amazon CEO Jeff Bezos responded to this as a testing-stage experiment of showing differential pricing to different customers. He explained that it had nothing to do with customer data and finally ceased the experiment.

Today, twenty years later, "big data rip-off" has become a common negative phenomenon. According to the survey results released by Beijing Consumer Association in March 2019, 88.32% of the respondents believed that "big data rip-off" was common or very common, and 56.92% of respondents said they had been ripped off by big data before.

A Harvard Business School study found that the average rent charged by a non-black landlord on airbnb.com is $144 per night while the rent of a black landlord is $107 per night. US retailer Staples uses the algorithm to implement the "one place, one price" policy that makes the discount in high-income areas even greater than that in low-income areas. Price discrimination is making other forms of discrimination legal. The fear of "crowd fishing" in daily consumption activities leads to a severe trust crisis among sellers and consumers.

More and more examples show that the objective algorithm discrimination and biases will make the solidification trend of social structure more obvious. Early in the 1980s, St George's Hospital Medical School in London used computers to browse enrollment résumés and preliminarily screen applicants. However, after four years of operation, it was found that this program would ignore the applicant's academic achievements and directly reject applicants who were female or without European names. This was the earliest case of algorithm bias against gender and racial differences.

We continue to see similar cases even today. Amazon's one-day delivery service does not cover black communities. The proportion of black people being wrongly marked by Correctional Offender Management Profiling for Alternative Sanctions (COMPAS)—the algorithm used by the U.S. State Government to assess the defendant's risk of recidivism—was disclosed to be twice that of white people. Many people keep being excluded from their favorite jobs due to algorithm automatic decision-making. To make the matter worse, the algorithm automation decision-making is neither open nor questioned. Thus there is no means the person concerned can access the cause for the result nor offer any correction.

In the face of opaque, unregulated, controversial, and even wrong automatic decision-making algorithms, the biases and injustice caused by "algorithm discrimination" or "algorithm tyranny" will become more and more inevitable. As algorithm decision-making penetrates into our lives, there is no way we can avoid having our personal data such as work performance, development potential, solvency, demand preference, health status, and others being collected, tracked, and controlled by algorithm users.

Algorithm automatic decision-making is everywhere, in loan limit determinations, recruitment screening, policy making, and even judicial assisted sentencing. The original structure of society will be further solidified, and the flow of individuals or resources outside the structural framework will be increasingly limited. The algorithm's accurate evaluation of the costs and rewards of each subject's relevant action will make some of them lose the opportunity to obtain new resources. While it seems to reduce the risk of the decision-maker, it may lead to injustice to those evaluated.

- **How to Deal with "Big data Rip-off?"**

Despite the fact that consumers may clearly know when they are being ripped off, few would choose to speak up for their rights. Research shows that only 26.72% of the respondents choose to report to the market surveillance department. 19.84% choose to speak with the seller directly or seek support from public media. The majority 53.44% took no action.

In fact, the reason for remaining silent is the difficulty in speaking up. The uncertainty with data behavior regulations, including data ownership, definition of major responsibility-holder, and legal boundaries for data competitions leaves ample grey areas for "big data rip-off." In order to mediate this dilemma, we will need multimodal reform in legal regulation, business ethics, and consumer's self-protection awareness development.

In terms of legal regulation, it is necessary to optimize the current regulatory system to be more pertinent and efficient. Since price frauds such as "big data rip-off" are usually massive in scale, implicit, and various which add up to the difficulty of identifying illegal operators on the edge of infringement in practice, a clear constitutional definition for the essence and peripheral meaning of "big data rip-off" and subsequent regulation needs to be made. This includes making penalty systems, increasing illegal costs, and making it a mandatory requirement for online platforms to demonstrate its service agreement and payment rules.

Sellers should be responsible of providing evidence for not implementing price discrimination, and this should be indicated in the Law for the protection of consumers' interests. The process for consumers to legally protect their rights should be simplified, and the public education on rights-protection awareness, means, and procedures needs to be further improved through public media and community lectures.

There also needs to be big data surveillance platforms to monitor ripping-off behaviors. These platforms will decide whether the enterprise is liable for implementing price discrimination through data analysis and reflecting the results to the consumers. In addition, a governmental department that inspects, monitors, and regulates big data information development should be established. It will help reinforce control over big data development, increase control efficiency, strengthen pertinence and overall planning, shorten the problem solving period, and provide the public with the means to report.

In terms of business ethics that advocates enterprise self-regulation and "user portrait" standardization, it is an essential supplement for legal regulation especially when the latter is insufficient, as is the case of China at present. The question of how to define Internet business ethics, making it the "trade conventions" that prioritizes consumer privacy protection and encourages enterprise self-regulation is already imminent.

Right now, each subfield in the market has its own self-regulated items. The big data society requires them to apply these rules to the Internet businesses level. For example, in travel services, the relatively well-developed regulations for taxis are also applicable for "web cabs" with additional items on consumer's privacy protection.

The role of trade associations as self-established organizations for protecting the interest of enterprises is also crucial here in setting a good market access and exit mechanism.

Finally, consumers need to improve their self-protection awareness and ability. There are a lot of "traps" in the world of the Internet, such as the "P2P thunderstorm" incidents that happen again and again. Astonishing reports on scams created through the Internet are also not rare.

Over-reliance on big data technology can result in the loss of subjective initiative and the servitude state of "information cocoons" as the individual is being controlled by big data.

Consumers need to develop their awareness in critically viewing the role of big data technology in human society. In addition, they need to pay attention to keeping a balance between the real and virtual worlds.

• Data Regulation for Living a Digital Life

Living in a digital world, we are "digital" before we are "social" or "economic." We are recorded, expressed, simulated, processed, and predicted by data, and so is the discrimination in real space.

Therefore, data regulation should come before algorithm regulation, and this requires government level management as well as guidance of individual and group behavior. At the same time, technical and constitutional supports are also indispensable.

The EU *General Data Protection Regulation* (GDPR) that came into force on May 25, 2018, further strengthened the protection of natural personal data on the basis of the 1995 *Directive 95/46/EC*. The GDPR not only provides a series of concrete legal rules but also conveys the idea of "data justice" in addition to "data efficiency." This document guided the path of big data development in China.

First, an individual's right to choose should be respected. When automated decisions may cause legal consequences or similar effects on individuals, the person concerned should be exempt from relevant restrictions unless he or she expressly agrees or it is essential for the conclusion and performance of contracts between the parties.

Second, personal sensitive data should be excluded from AI automatic decision-making. This includes data regarding political orientation, religious beliefs, health conditions, sex life, sexual orientation, or genetic data and biological data that can uniquely identify natural persons. Because the disclosure, modification, and improper use of these data will have a negative impact on individuals, it is important for lawmakers to provide for more careful and responsible collection, use, and sharing of any sensitive data that may lead to discrimination.

Third, data should be transparent in order to identify and challenge discrimination and bias in data application. In other words, an individual should know more about the data to repair information symmetry under the increasingly complex situation of data production and processing. For example, banks should inform the lenders that AI may be used to audit their qualifications, and the worst results of the audit (such as not approving loans) should also be stated in advance. The "useful information" listed should not only be comprehensible to everyone but also helpful for individuals to claim their legal rights.

Algorithm regulation needs to enforce the technical standards and traceability of algorithms. The current algorithm is essentially a programming technology, and the most direct way to regulate it is to formulate standards, which is also the most direct basis for management carried out by relevant national departments.

Basically, we should comprehensively improve our standard understanding of AI algorithms, improve the predictability of new industrial system costs, and reduce the chaotic development of new technology. At the international level, the technical specification system is usually expressed as a system of "technical regulations + technical standards + technical certification."

Among them, technical regulations are mandatory requirements for the descriptive technical features of different industries. Technical standards are requirements for specific technical indicators. Their main function is to support technical regulations. In China, the effectiveness of technical standards corresponds to group standards. However, these standards have no legal effect in China. This calls for standard awareness and investment in government, enterprises, and society.

In the meantime, we should coordinate the forms of responsibility stipulated in the existing legal system. According to the *E-Commerce Law*, the *Network Security Law*, and the *Regulations on Computer Information Systems Security Protection*, there are two main ways to regulate algorithms: one is to directly regulate algorithm programmers and algorithm service providers. For example, the *Network Security Law* clearly stipulates that network services should comply with the national standards and should not set malicious programs. In case of any problem, it is necessary to remedy the loophole and ensure safety in time.

The other is to clarify the regulatory responsibilities of relevant administrative departments to indirectly curb the occurrence of algorithm infringement. For example, the *E-Commerce Law* mainly regulates the behavior of e-commerce operators and sets their relevant responsibilities, including regulations on information disclosure and protection, searching and advertising, services, and transactions. The *Network Security Law* mainly stipulates administrative organs' responsibilities in monitoring, maintaining, and managing, including supervising, early warnings, emergency disposal, network operations, and information security maintenance.

Of course, the change of any social rules is accompanied by technology development. In the face of rapidly changing technological challenges, what we can do is to incorporate algorithms into the rule of law so as to create a more harmonious era of big data.

5.2 AI Security Competition

History shows that network security threats increase with new technological progress.

Relational databases bring SQL injection attacks, Web scripting language encourages cross-site scripting attacks, and IoT devices open up a new method to create Botnets. A Pandora's box has been opened, and the disadvantages of digital security have been revealed. Social media has created a new method to manipulate people through micro-target content distribution, and it is easier to receive information about phishing attacks. Bitcoin makes it possible to encrypt ransomware blackmail software attacks.

As new attack methods continuously emerge, and security vulnerabilities and malware multiply. In 2019, VulnDB and CVE recorded more than 15,000 security vulnerabilities, with an average of more than 1,200 per month. The same year, CNCERT captured more than 62 million computer malware samples, spreading more than 8.24 million times a day from more than 660,000 computer malware families.

According to the data of research group IDC, the number of networked devices is expected to reach 42 billion by 2025. In view of this, our society is entering the era of "super data" which introduces a new round of security threats at a time when data algorithms are popular and AI technology is in the ascendant.

- **How to Realize AI Attacks?**

Let's first try to imagine this:

The future terrorists will not need bombs, uranium, or biological weapons to carry out an attack. The only things they will need are some adhesive tape and a pair of sports shoes. By adding a piece of adhesive tape on the traffic lights at crossroads, the terrorists can make the automatic driving vehicles recognize a red light as green and cause car accidents. This is enough to paralyze the transportation system at a major intersection with large traffic flow in the city, and it only costs $1.50.

This is what we call an "AI attack." But how is it achieved?

To understand the uniqueness of AI attacks, we need to first understand the concept of AI Deep Learning. DL, which is a sub-field in machine learning, indicates software checking and comparing large amounts of data to create logic of its own. While machine learning has existed for a long time, DL has only gained popularity in the recent few years.

An artificial neural network is the basic structure of a DL algorithm that roughly imitates the physical structure of human brains. Contrary to traditional software development that requires programmers to write rules that define the behavior of applications, the neural networks create their own behavioral rules by learning from a large number of examples.

When a training example is provided, a neural network runs it through a layer of artificial neurons and then adjusts their internal parameters to be able to classify future data with similar attributes, which is very difficult for manual coding software.

As a simple example, if you train a neural network with sample images of cats and dogs, it will be able to tell you whether the new image contains cats or dogs. It is very difficult to perform such tasks using classic Machine Learning or older AI techniques. It is generally slow and error-prone. Computer vision, speech recognition, speech-to-text, and facial recognition that have emerged in recent years have all made great progress due to DL.

Yet because the neural network relies too much on data, it can sometimes make mistakes that seem completely illogical or even stupid to humans. For example, in 2018, the AI software used by the Metropolitan Police in the UK to detect and mark pictures of child abuse incorrectly marked pictures of sand dunes as nude.

When these errors occur, the "DL method" that AI algorithms are proudly based on will turn against them into a weapon that enemies can apply to attack and manipulate. Therefore, what appear to us as a slightly defaced red light may actually appear green to the AI system. This is called an adversarial attack of AI, which means the wrong irrational input that emphasizes the fundamental difference between the functions of DL and human thinking.

Similar errors that can possibly be exploited by terrorists are imminent. Not long ago, a group of elementary students cracked the face-screening function of the Hive Box smart express cabinet simply with a printed photo of their parents. They successfully tricked the "intelligent" express cabinet to take out their parents' parcels. In addition, two teenagers from the University of Leuven in Belgium easily escaped the surveillance of YOLOv2, an outstanding example in target detection devices, and became invisible just by placing a picture in front of their bodies.

With the development of AI technology, there will be more times in our lives when we will need this kind of biometric technology. It means that once it is attacked, it will cause endless harm. The world-renowned media The Next Web (TNW) also reported earlier that hackers can deceive the voice-to-text system in certain ways, such as secretly adding some voice command to the user's favorite songs that can make the smart voice assistant to transfer the user's account balance. In addition, adversarial attacks can also deceive a GPS to mislead ships and autonomous vehicles, and even modify AI-driven missile targets.

- **DL-based Cyber Threats**

While the global digital age has just begun, hacker attacks are already widespread. In 2007, the Worm.Whboy virus ravaged the Chinese Internet; in 2008, the Conficker worm infected tens of millions of computers; in 2010, Baidu was attacked by the worst hacker in history; in 2014, Sony Pictures was attacked to the point the chairman had to resign; in 2015, The US government was attacked that caused employee data disclosure.

When the research in AI technology gained popularity, the conflicts that took place in cyberspace also leveled up. Predicting the possibility of hackers manipulating AI technology to enact attacks may help us achieve better results in protecting the online world.

At present, most of the malicious software of cyber threats is generated manually. That is, hackers will write scripts to generate computer viruses and trojan horses, and use Rootkits, password capture, and other tools to assist in distribution and execution.

How does machine learning assist in generating malware?

When machines learn from data retrieved from malware samples (such as title fields, instruction sequences, and even raw bytes), they can establish a model for distinguishing benign and malicious software. However, analysis of security intelligence reveals that machine learning and in-depth neural networks can be confused by evasion attacks (adversarial examples).

In 2017, the first example of publicly using machine learning to create malware was proposed in the paper *Generating Adversarial Malware Examples for Black-Box Attacks Based on GAN*. Malware authors usually do not have access to the detailed structure and parameters of the machine learning model used by the malware detection system, so they can only perform black box attacks. The paper demonstrates how to build a generative adversarial network (GAN) to generate anti-malware samples that can bypass the black box detection system based on machine learning.

If the AI of network security companies can learn to identify potential malware, hackers can make decisions by observing and learning anti-malware software and use this knowledge to develop malware that is "minimally detected."

At the 2017 DEFCON conference, the security company Endgame revealed how to use Elon Musk's OpenAI framework to generate customized malware that cannot be detected by the security engine. Endgame's research is based on binary files that appear to be malicious. By changing part of the code, the changed code can evade detection by the anti-virus engine.

- **Data Poisoning**

Both AI adversarial attacks and hacker's DL-based malware escape belong to the category of an AI Input Attack, that is, manipulating the information input to the AI system to change the output. In essence, all AI systems are just a machine that includes input, calculation, and output. So the attacker can influence the end product by manipulating the entry.

Data poisoning is a typical poisoning attack, meaning covert manipulation during the creation of an AI system so that the system malfunctions in a manner preset by the attacker. This can be achieved because the only basis for AI to "learn" how to handle a task through DL is data. Therefore, if the data is contaminated with camouflage data and malicious samples that can destroy data integrity and lead to the training algorithm deviations in decision-making, the AI system will thus be polluted.

An example of data poisoning includes training facial recognition authentication systems to verify the identities of unauthorized persons. After Apple launched a new neural network-based Face ID authentication technology in 2018, many users began to test its range of functions. As Apple has warned, in some cases, the technology fails to tell the difference between identical twins.

But one of the interesting failures is the situation of two brothers. They are not twins, they look different, and they differ in age by many years. The brothers initially released a video showing how to unlock an iPhone X with Face ID. But then they released an update in which they showed that they actually cheated the Face ID by training their neural network with their face. Of course, this is a harmless example, but it is easy to see how the same pattern can serve malicious purposes.

This is also mentioned in the *AI Data Security White Paper* (2019) issued by the China Academy of Information and Communications Technology. It pointed out that the data security risks for AI include: training data pollution leads to AI decision-making errors, data abnormalities in the operation phase lead to errors in the operation of the intelligent system (such as adversarial sample attacks), model theft attacks that reverse the data of the algorithm model, etc.

It is worth noting that with the in-depth integration of AI and the real economy, launching cyber attacks in the training sample becomes the most direct and effective method for constructing data sets in the medical, transportation, and financial industries. The potential harm is huge. For example, in the military field, the use of information camouflage can cause devastating outcomes by inducing autonomous weapons to start or attack.

- **Attack and Defense in the Era of AI**

In the approaching "time of machines," network security will be a huge issue that involves in-depth global connections and cooperation. It will be characterized by the common duty of humans and AI. With the explosive growth of various Internet technologies, the means of network attacks are constantly enriched and upgraded. Defending against network attacks takes the ability to quickly identify, respond, and learn.

It is undoubtedly difficult to detect a virus threat intrusion with machine learning. We can only possibly achieve breakthroughs in cyber security under the comprehensive application of technology, with an understanding of the information leakage and its correlation (i.e., how the hacker invades the system, what is the path of the attack, which link appears problematic), or with an analysis of the causal relationship graph to increase the interpretability from the analysis end.

It also needs intelligent dynamic defense capabilities, as the essence of network security is the confrontation between offense and defense. In the traditional offensive and defensive model, network attackers usually take the initiative, and the security defense forces can only passively accept the attack. However, in the future security ecosystem, members can stimulate each other through data and technology exchanges and information sharing, automatically upgrading their security defense capabilities, and even predict threats to a certain extent.

Network security is a highly confrontational and dynamic field by nature. It opened up for the field of anti-virus software a "blue ocean market" where the AI anti-virus industry is offered major development opportunities. The anti-virus software industry should first have the awareness of preventing AI viruses, and then it should pay attention to information security and functional safety issues in terms of software technology and algorithm security.

5.3 Sex, Love, and Robots

With the advent of the era of AI, gendered robots are developing both in terms of technology and demand.

Technically, AI technology has become more sophisticated, allowing more and more realistic, and widely celebrated gendered robots to enter the market and become an important component of the sex industry. Nowadays, the most mature gendered robots cannot only acquire knowledge through AI and generate emotions with humans, but they can also have different personalities and have enough emotional models for users to choose from.

In China, there are 240 million people who are currently single; among those who are married, there are 290 million who are sexually disabled and 80 million who are hyper-sexual. BI

(Business Insider) news shows that since the outbreak of the 2019 pandemic, sales of Harmony, the world's first AI gendered robot, have surged, in spite of the high price of $12,000.

In 2019, futurist Dr. Ian Pearson published a report on the prediction of future sex. Pearson believed that around 2050, sex between humans and robots will become very popular, and robots may even replace human sexual partners.

Dr. Pearson's prediction seems to be becoming a reality. As the cost of gendered robots continues to drop and their functions continue to improve, it is not long before we can live the future where gendered robots are more diverse and sexier.

- **Is the future of gendered robots here already?**

In 2018, the world's first AI robot, Harmony, was officially put on sale at a price of £7,775, which is approximately RMB 68,000.

Harmony demonstrated an extremely high level of performance. It has more than 30 different faces including that of Black people and Asians. Designers will polish these faces by themselves and even spray the freckles one by one to ensure a perfect appearance.

In addition to highly simulated human body shape and flexibility, Harmony can acquire knowledge through AI technology and generate emotions with humans. It has 12 different personalities (i.e., kind, sexy, innocent). Any Harmony robot has its own exclusive app. It can be connected to the Internet through the app and communicate with people using the corpus.

Compared to the first generation, the upgraded Harmony 2.0 that soon came out was more like a real companion. It had more facial expressions, more flexible limbs, and a more realistic-feeling skin. The internal heater also allows it to simulate real body temperature.

What's more, Harmony 2.0 integrated Amazon's Alexa voice system that enabled it to analyze the voice information in a timely manner, and give feedback quickly and accurately on various topics. At the same time, Harmony 2.0 was also equipped with an intelligent software system that can store and memorize the content of past chats to determine the habits and preferences of the partner. In other words, the robot will "understand" its human partner more and more as the time they spend together accumulates.

The epidemic has further catalyzed the demand for gendered robots. Data shows that during the large-scale outbreak of the epidemic, Harmony's market sales increased by at least 50%. Silicone Lovers also stated that the company has been receiving orders for gendered robots more than ever.

The reason for this trend is as follows: under the epidemic, people have less contact and communication and thus more need for companionship. Although the price of current sex robots is still beyond the reach of most people, it is foreseeable that the commercialization and normalization of gendered robots is no longer far away.

Of course, robots do not have sexual desire, and the complete form of the gendered robot does not exist yet in 2020. But it is safe to say that more personalized, precise, and better sexual experiences will be available very soon.

For example, VR devices can create avatars of gendered robots, and biochemical signal systems can directly release pheromones that form sexual stimuli to the human body. The brain-computer interface may create connections between the app and the human brain to directly send electrical signals of sexual pleasure and achieve hand-free orgasms.

• Sexuality and Emotion, Companionship, and Marriage

In addition to sexual desire, humans also have emotional needs. This is also a heated topic about gendered robots.

Humans are social animals, but not everyone can gain solace from human companionship. In fact, the intervention of science and technology in modern society is creating more loneliness and making human emotional needs a larger market. Therefore, we are curious to find out if gendered robots can solve this problem.

The "virtual boyfriend" test initiated by Microsoft Xiaoice attracted 1.18 million human "girlfriends" just seven days after it went online. They would do sweet talk and share their living habits with the software, and it could make them happy only by giving mechanical answers according to recognized key words.

Li Zeyan, one of the male protagonists of the mobile game "Love and the Producer," won himself seven million "wives" in one month and reaped RMB 300 million from them. What the AI character did was simply following the script and played its part like a real human being.

In this way, gendered robots can indeed meet the emotional needs of users. For involuntary single people—those who cannot find a partner due to different involuntary reasons, such as widowed elderly, disabled people, and people with sexual dysfunction in their later years—gendered robots can assist them in solving sexual problems and provide them with emotional support.

But if we look further, we will find that silicon-based bisexual robots cannot generate and use emotions like carbon-based creatures do. AI can train its data according to behavioral input and imitate human beings, or to "fake" subjective experience and deconstruct human "loneliness and love." But while gender robots may be able to replicate people's emotions, they cannot after all create emotions.

Any robot learning will need an established goal. Even for algorithms as advanced as the AlphaGo, it can only fabricate experiences through the goal and data that scientists provide it. However, it is difficult to define goals for emotional needs due to its unpredictable randomness and irrational originality.

In fact, it is the possibility of achieving perfection that makes machines and algorithms inferior to human. All perfections are similar and even boring, but it is imperfection that makes us human.

With gendered robots, people can gain the pleasure of control, but they may also lose the pleasure of losing control at the same time. We are guaranteed absolute security, but we are also giving up the sense of uncertainty of being in an exclusive relationship. The reason that gendered robots cannot replace humans is that they have no right not to love.

When bionic gendered robots enter human society, they will inevitably bring huge changes to marriage. Marriage may not necessarily disappear completely, but it is bound to become more diversified.

When gendered robots can eventually replace human partners, the "symbiotic relationship" between men and women will become a "competitive relationship." Perhaps the concept of "family" will no longer exist, because every family composed of both spouses will become independent "individuals." Perhaps the concept of human beings as the sum of social relations may no longer exist, because "family" by then will be individuals with the same social competitiveness and their artificial partners.

In the meantime, people's sexual desires will be greatly reduced because customized artificial partners can "absolutely obey" any needs of their users. Sex, as a sole human behavior, is diverse. It is still unclear whether the gendered robot will eliminate people's curiosity in sexual exploration due to mechanical limitations.

The popularization of gendered robots also faces more questions regarding social ethics, including the legal regulation of material safety, data safety, sexual objectification and violence against women, and other questions.

5.4 The "Ethical Vacuum" of AI

AI is not merely a revolution in the Internet field or a disruptive technology in one specific industry but an important tool that supports the entire industrial structure and economic ecosystem. Its energy can be projected in almost all industry sectors, promote the transformation of its industrial form, and provide new momentum for global economic growth and development.

Throughout history, no technology has ever triggered human imagination like AI. While beneficial to the human society, it has also become a prominent international scientific controversy. The subversive nature of AI technology makes us rethink the huge dangers hidden behind it. As early as in November 2016, the *Global Risk Report* compiled by the World Economic Forum listed 12 emerging technologies that were in urgent need of proper management. AI and robotics topped the list.

• Finding Spiritual Sustenance for AI

The connotation of the term "intelligence" can literally cover almost all human activities. Even within the domain of AI technology alone, what AI can achieve is far beyond people's imagination.

Our emails are selected for us through a spam filter, and our taxicabs are called for us through cab-hailing apps. The news we see every day is recommended for us based on AI algorithm computing, and the homepage of our shopping websites displays products we are most likely to buy. More and more simplified AI vehicles and systems in our daily lives, some deeply felt by people while others less obviously observed, are assisting social development but at the same time creating hidden dangers.

In general, the subject of AI management consists of two parts. One is the governance of the technology itself, and the other is the standardized governance of derivative issues.

Research in the field of intelligent robots has continued for 40–50 years. Not only primitive industrial robots, but service robots also have a research history of more than 30 years. These studies have greatly enhanced the operational capabilities of machines, allowing machines to replace and even surpass humans in various activities.

In the past five or six years, machine learning algorithms have developed rapidly. Although there is still a long way to go before machines can fully attain "comprehension," the emergence of better and cheaper hardware and sensors, as well as wireless low-latency interconnection between devices and continuous data input are suggesting a larger expansion capacity for the perception, understanding, and networking capabilities of machines.

Compared with human intelligence, which has slow basic components, a large number of non-modifiable primitive instincts in structural coding, and limited acquired ability to self-plasticize, AI has a far greater future development potential. Although it is impossible to determine the exact time point as we can with singularity theory, there is no doubt that machine intelligence will eventually surpass human intelligence.

AI governance requires us to jump out of the realm of technology and establish a mind for the machine from a humanistic perspective, including the values and morality of scientists in related fields in the development and application of AI technology.

For a new technology, it is usually the application of it that causes harm rather than the technology per se. The same is true for AI. All the tools that we have learned to use within the past several decades—computers, the Internet, Cloud Computing, AI, and 5G communication technology—help us expand the breadth of information and the search for knowledge, deepen the depth of reasoning, and do very complex calculations. How to better master and use these technologies to help humans maximize their intelligence potential is the key to the problem of AI management.

• The Derivation Problems Faced by AI Technology

The standardized management of AI derivation problems is another crucial aspect of future AI development in addition to its crisis management.

First of all, the issue of employment is the closest to people's livelihood security and the most in need of a solution. Due to the profit-seeking nature of capital, in an era where science and technology are the primary productive forces, AI's huge productivity will definitely attract capital accumulation, and the development of the digital economy will further promote the automation and digital transformation of enterprises.

As a result, automation reduces the horizontal division of labor in society, and AI will achieve semi-automated and fully automated production, thereby eliminating the dependence of production on individual human differences. With the deepening of AI research and development, AI has taken up the more dominant role in the social division of labor. The technological differences among ordinary workers gradually disappear.

However, the increasing possibility of AI substituting human laborers in the workforce does not lead to better living conditions for ordinary workers. On the contrary, as workers depend more on technology, their own competency in the division of labor system will decline.

In addition to reducing the horizontal division of labor in society, AI technology also monopolizes the vertical division. In the information age, automated production based on big data and computer applications has become the main driving force for economic growth. Huge amounts of capital have poured into enterprises that master cutting-edge technology.

One result of this trend is that some routine jobs will be replaced by machines and the unemployment rate will go up. At the same time, companies that are already at the upper level of the social division of labor will become more consolidated in their status due to the large inflow of capital. They will even cut off other companies and laborers' opportunities to break through the social division of labor and strengthen the existing, unequal hierarchical order. If the fracture zone between the top and bottom of the social division of labor becomes insurmountable, the social polarization will become extreme.

Data management is another thorny problem with AI development and application. When information regarding individual location, content of interest, and itinerary arrangement are readily available, it means that people's privacy will be widely open to the public attention. The dynamic monitoring of big data makes exploring and collecting "traces" of individual user operations possible. With the help of AI, we cannot only understand "who" you are but also predict "who" you will become.

Although big data and other intelligent methods provide great convenience to group classification, they also make people more vulnerable in terms of privacy protection. Over-reliance on data algorithms and blind expansion of the scope of applications thus raise ethical concerns.

Finally, the development of AI technology poses challenges to international relations. As we can see from past experiences, each World War and Industrial Revolution had rapidly reconstructed the social order and international patterns. It is inferable that the fourth technological revolution represented by AI technology we are witnessing right now will have a disruptive impact.

In September 2018, the McKinsey Global Institute's special report on the impact of AI on the world economy showed that by 2030, AI may contribute an additional US$13 trillion in global GDP growth and promote GDP annual growth by approximately $120 million. In this new round of industrial revolution, whoever masters the main power of AI technology is likely to occupy the lifeline of the global economy in the next few years with a monopoly position.

Like nuclear power, AI technology can both be used in industries and for military purposes. Many countries have begun to increase strategic deployment of AI in the military field, especially in the research and development of autonomous weapons. When AI technology gives the country more political and military power, it also exacerbates the security dilemma and the challenges to AI governance for the international community.

The truth is, we have entered a special zone of modern "ethical vacuum" which is caused by the lack of traditional ethics during the development of modern natural sciences. Human behavior and thinking patterns are rapidly moving forward and are irreversible, while the development of traditional ethics is slow and lagging. An ethical vacuum is thus created.

To sum up, vigorous AI development must be accompanied by studies of AI ethics so that effective AI management can give full play to the true value of the new technologies.

5.5 When We Talk About Facial Recognition

In 1964, Woody Bledsoe proposed the world's first facial recognition algorithm that used chain codes as an identification feature.

In the 1970s, a 2D face recognition algorithm was born. It gave rise to a new recognition which prepared the 3D facial recognition algorithm for higher efficiency and accuracy.

At present, facial recognition technology has been embedded in all aspects of social production and people's daily lives. It is widely used in fields such as financial behavior, workplace supervision, safety prevention and control, and so on. From 2015 to 2019, the number of patent applications for facial recognition and video surveillance soared from 1,000 to 3,000, 75% of which were in China. According to the market research firm MarketstandMarkets, the global facial recognition market will reach $7 billion by 2024.

But facial recognition technology is not well received by the public. In 2019, ADA Lovelace Institute found that 55% of the respondents who received their survey on facial recognition wanted the government to restrict the use of this technology by the police. Respondents also felt uncomfortable about its commercial use. Only 17% of the respondents wanted to see facial recognition technology used for age verification in supermarkets, 7% agreed to use it to track customers, and 4% thought it appropriate to use it to screen job seekers.

When discussion on facial recognition centers on the divergence of technology and public opinion, what exactly are we talking about when we talk about facial recognition?

• Privacy Costs in Facial Recognition

Since big data is the foundation both for the rapid development of AI and the Internet with 5G network, people's behavior data get to be preserved to a large extent. When this data information accumulates in the memory of the Internet, it becomes a tool to monitor people's behavior. "Privacy" will become a term of the past, because people are forced to be "transparent" in the era of big data.

In the facial recognition scenario, users are not only giving out their facial geometric features but also sensitive information about their age, gender, emotional features, and others.

Kantar Millward Brown, a market research firm, applied the technology developed by Affectiva, an American start-up, to evaluate consumer response to TV advertising. The app records people's facial information and their expressions frame by frame to evaluate their emotions.

The company's director manager of the Innovative Department Graham Page said that by monitoring people's expressions, we can obtain richer information and even accurately see which part of the advertisement works and what kind of emotional response it evokes.

Related technologies have become more and more dependable within the context of AI and DL. If facial information can be connected with interest, personality, consumption habits, and even whereabouts on the Internet, we can thus delineate a clearer information portrait of the individual. But this growing data self in the memory of the Internet has also become a huge safety hazard against individual privacy.

Online data with a personal "profile" can extend wirelessly in cyberspace; offline, the combination of ubiquitous cameras and facial recognition applications leave individual activities in a highly monitored environment. People are now fully dealing with the dilemma of "privacy streaking," and it has become extremely difficult to protect personal privacy.

On the other hand, while facial recognition technology has penetrated all aspects of our life, issues with subsequent storage and use of facial information is still largely unsolved.

It's now been a year since the "first case of facial recognition." Guo Bing was an annual card-carrying member of the World Wildlife Fund (WWF). In October 2019, he received a short notice saying "the previous fingerprint recognition admission system has been cancelled and annual card members are now admitted based on facial recognition. From now on, users without registered face recognition will not be able to enter the park normally."

Mr. Guo believed that facial information is personal sensitive information, and he requested a refund of the annual card fee for the World Wildlife Fund's unilateral and illegal initiative of modifying the terms of service. When his request was rejected, Mr. Guo filed a petition to bring the World Wildlife Fund to court. This is by no means an isolated case. Facial recognition is also seen being applied in subway security checks and in classrooms. In the past year alone, facial recognition has aroused controversy with high frequency.

There are two prominent issues with the improper application of facial information. First of all, the organization storing people's facial information is essentially operated by men, that is, a large number of facial information with strong identity directivity is controlled by a specific group of people. There's no way for us to know how these people will use our personal data and whether their operation is legal.

Second of all, facial recognition applications interpret and screen subjects through specific codes. The operation of the code is thus subjected to the possibility of being invaded by hackers. With the development of face forgery technology and the increasingly mature industrial chain of anti-real-name systems, it is possible to decode facial information and replace the "real face" with a "false face."

In other words, with the photo of a certain face and the facial features that can be recognized by the system, an AI algorithm can then carry out targeted training to copy the face image, including turning back and forth or blinking, so as to enable the corresponding service by using other people's facial information.

One AI Company in San Diego, USA, once successfully deceived many facial recognition payment systems, including WeChat and Alipay, using a specially made 3D mask. Although the two companies have both made an urgent response indicating that there were no other incidents of theft caused by similar technology, the maturity of 3D printing technology is obviously raising the risk of a facial recognition system being broken by "fake faces."

• From "Anonymous" to "Nonymous"

Our appearance plays an important role in social relations and personality development. For this reason, masked faces are often related to negative impressions and people who are untrustworthy or dangerous. Laws prohibiting masked faces in social gatherings are enacted in many countries and cities such as Germany, Italy, France, New York, and Hong Kong SAR, which also highlights the publicity of faces in public spaces.

But publicity does not mean full elimination of anonymity. Different from the traditional society composed of acquaintances, modern society composed of atomized individuals shows a stronger inclination towards anonymity. Although individuals completely reveal their faces, whereabouts, and speech, they still have the freedom of ignoring others. In subways, restaurants, streets, elevators, and other public spaces, it is a shared social norm to "politely not pay attention to others."

However, AI, facial recognition, and other emerging technologies have pushed us into a new era of "digital human rights," which is compatible with both positive and negative aspects that challenges "public strangeness" and anonymity in modern society.

The positive orientation of digital human rights implies that the state should take the initiative in the promotion and realization of digital human rights. In an environment where people can hardly escape from living a networked lifestyle, the Internet becomes an essential infrastructure for the public like transportation, electricity, and tap water.

Therefore, the state is responsible for building the Internet infrastructure, completing hardware and software engineering construction, and providing "Internet plus" public services.

The negative orientation of digital human rights means people's "right to be alone" in the era of big data. When any individual is connected with the Internet and expanding her own living space, she should also be protected from examination, spying, and identification when her identity has nothing to do with national and social security, interference with her way of living, and infringement of her personal interests. In addition, she should also be allowed to improve her ability to do whatever she wants with the premise of the interests of the state, society, and others.

This "right to be alone" enables individuals to enjoy the "negative freedom" of not being interfered with and to develop their unique personalities. It ensures the tolerance and diversity of the society and protects those who look different (i.e., ethnic minorities, foreigners, disabled people) or behave deviant from discrimination.

Even, as German sociologist Armin Nassehi said, this is the only way that the society can bear the inequality and injustice caused by social transformation because "It depends on invisibility rather than visibility; strangeness rather than intimacy; distance rather than closeness." In this regard, "social solidarity is based on strangeness."

However, the increasing number of cameras and facial recognition driven by algorithms and big data is forcing people to move from being "anonymous" to "nonymous." The sense of strangeness has disappeared, but the sense of intimacy and security of an acquaintance society hasn't returned.

On the contrary, facial recognition applications may discriminate against specific populations more easily. For example, groups with distinct facial features or other information that can be detected through facial features can make them focus of attention. This is because facial recognition apps based on any algorithm depend on big data, which is not neutral.

Studies have shown facial recognition is biased against racial differences. In airports and train stations, some people may not be recognized normally due to an algorithm bias and must be interrogated and checked by staff. In addition, discrimination may also exist in subsequent services.

Therefore, in the "face scanning" era, our faces that were once the basis for interpersonal communication and trust building are now tools for recognition, that can be captured by cameras and other devices to generate digital images to identify, authenticate, and verify one individual from another. This process devalues our personality and dignity and turns an independent human being into a series of numbers and symbols. This is the real crisis that facial recognition entails.

When considering facial recognition technology, we should not only consider what is legal but also what is moral. Right now, facial recognition has brought severe challenges to social governance, such as the important personal information involved in its application and the realization of digital human rights. All these suggest that we should mind the legitimate boundaries of facial recognition when applying the technology.

5.6 Robot Neighbors

Accepted or not, AI is now deeply integrated with our life. On the one hand, it brings efficiency and convenience and promotes the sustainable development of a social economy; On the other hand, it brings challenges to existing ethical cognition.

As the broadest field of intelligent robot development in the future, "Robot Companions" are more and more involved in people's lives as assistants, friends, partners, and even family members. When we enter the era of man-machine coexistence and become neighbors with robots, how should we get along with them? When robots are more and more endowed with human emotions, what should we do in response to the new relationship between human and machines in addition to that between people and society?

• How Do Humans Get Along with Robots?

Artificial intelligence, as its name suggests, is a machine's functional simulation of human intelligence. AI technologies are fundamentally different from any technology in the past agricultural age or industrial age, and they have completely redefined in every possible aspect the way human beings connect with the world.

In an industrial society, people's awe of technology is natural and prominent. Technology is regarded as a tool with class and power attributes. People who master technology are usually granted higher power and maintain a higher social status. In the era of AI, intelligent technology covers and is integrated in people's lives. Therefore, rediscovering the relationship between man and technology has become the basic paradigm of current research.

For all kinds of accompanying robots, whether invisible intelligent software, intelligent speakers, or humanoid robots that may appear in the future, they are agents or actors that can interact with people in meaningful ways. "Robot" becomes a subject or entity related to a human "image" that can interact with people rather than merely being a tool or machine.

In fact, any entity with a human appearance or can interact with people obtains a special meaning. With statues of historical or artistic figures, for example, when facing them, our hearts are often filled with some kind of reverence or cordial emotion.

The highly developed intelligent robot will not only have the appearance but also gradually the attributes of human speech and behavior in line with human etiquette, rapid reasoning and thinking ability, compliance with human legal and moral principles, etc.

So how do we get along with robots? Should we treat accompanying robots with mutual equality and respect like we do with a human counterpart? Or can we give vent to our anger whenever we feel it and in whatever means we desire?

At a Science Fair in Austria at the end of 2017, Samantha, a gendered robot developed by Dr. Serge Santos, was repeatedly molested and violently treated to the extent that two of her fingers were broken. Until now, one of the issues with intelligent robots is that robots are unconscious and do not realize that they are robots. Even if they are endowed with human appearances and attributes, their behaviors are still the results of calculation after all. So, no matter how people treat them, they will not feel happiness or pain like a real human do. The happiness and sadness they present are designed for the users. The question of "whether humans can treat robots in whatever ways they like" does not directly involve ethical concerns, but it calls for our reflection in the moral aspect.

Although the factual basis for these moral restrictions is still not completely clear, it is safe to say that robots based on strong AI technology and genetic engineering technology can enjoy a moral status as moral acceptors. In fact, as early as 2017, the Saudi Arabian government officially granted the robot Sophia citizenship in the Kingdom of Saudi Arabia. For more gendered robots in the future, their chances of playing an important role in people's life are greater.

Necessary normative research should be carried out in order to avoid technology abuse. We should respect not only our own human nature but also the human nature reflected in robots.

- **Establishing Legislation for Robots**

Current accompanying robots are still in their primary developmental stage. How to design robots before people interact with them is a major issue. A robot is not a simple commodity in the traditional sense. When the robot displays a human appearance and attributes, it is important that we establish specific legislation for the robot industry.

People are often affected by the behaviors of machines when interacting with them. This influence is usually invisible but real. Therefore, robots should be designed to communicate with human beings according to human rules.

For example, as an important subdivision of future companion robots, sex robots cannot be designed simply as sex tools. The program should also incorporate some basic rules of interpersonal sex communication (i.e., the principle of informed consent and the principle of mutual acceptance).

Another example is the design of soldier robots and police robots. We need to avoid designing them only as military robots because their moral judgments and actions directly concern human lives. Therefore, their design and production not only need to be transparent and open but should also be supervised by a fair-minded global organization. It is a must to incorporate the maintenance and observance of human core morality into their programs, and the design of such robots needs to follow common global standards.

In addition, although highly intelligent robots can enjoy the moral status as a moral acceptor, they can only become explicit moral actors at most instead of full moral actors like mature human individuals.

This means that robot designers need to bear some moral responsibility for robot behaviors. Therefore, they should try to restrict those designs with ulterior motives, like companion robots that deliberately lie. If robots could "learn" how to lie, it is inevitable that companion robots will shield those who are guilty and engage in illegal activities.

In general, humans are the moral guardians of robots. As we welcome our robot neighbors in our lives, we need to continuously reflect on our interactive modes. It is only through prudence and restraint can we refrain from self-abandonment and addiction with robots.

5.7 In-depth Forgery

AI makes everything in social life explicit and direct in the era of technology, but it also makes the impact of forgery deeper and long-lasting.

As a human body image synthesis technology based on AI, in-depth forgery was initially a simple idea for programmers to make funny "head-changing" videos. The superimposition of two DL algorithms finally created a complex system, which was expanded by AI progression. Examples such as high-quality facial-expression-matching, seamless switching, and face-changing-video generation have become available for everyone, from star politicians to any ordinary individual.

New developmental opportunities also bring security risks. As in-depth forgery becomes more complex but easier to make, it is challenging the domains of policy making, technologies, and legislation in a number of ways.

AI reshapes people's cognition, and people also transfer their inherent biases to the technology they are developing. What's more, people do not seem to be aware of all this. In fact, there is nothing they can do about the situation when the technology is cloaked in the concept of "entertainment."

• From In-depth Synthesis to In-depth Forgery

In the beginning, in-depth forgery existed as a form of AI technology of synthetic content. In-depth synthesis technology developed gradually out of the progress of a Generation Countermeasure Network (GAN).

GAN consists of two competing systems, the generator and the recognizer. The first step in establishing a GAN is to identify the required output and create a training data set for the generator. Once the generator starts creating acceptable output content, the video clip can be provided to the recognizer for authentication; if the video is identified as false, it will tell the generator what needs to be corrected when creating the next video.

According to the result of each "confrontation," the generator will adjust the parameters used in its production until the discriminator cannot distinguish the generated works and authentic works. It will superimpose the existing images and videos on the source image and finally generate the synthetic video.

Typical examples of in-depth synthesis are: face replacement, face reproduction, face synthesis, and speech synthesis.

Face replacement, also known as "face change," refers to "stitching" one person's face (source character) to another person's face (target character) so as to cover the target character's face.

Face reproduction uses in-depth synthesis technology to change facial features (i.e., inclination of the mouth, eyebrows, eyes, and head) in order to manipulate the target object's facial expression. Unlike face replacement, face reproduction focuses on changing someone's facial expression to create the impression that they are saying something that they have not actually said.

Face synthesis can create new facial images, and many of these randomly generated facial images can be comparable to or even replace some real portraits, such as in advertising, user avatars, etc.

Speech synthesis involves creating a specific sound model, which can convert text into a voice close to human intonation and rhythm. The Canadian speech synthesis system RealTalk is a good example of this system that generates a perfect human voice based on text input only.

The popularity of in-depth synthesis technology was fairly accidental. In 2017, user "deepfakes" uploaded some digitally tampered pornographic videos to the US news website Reddit. The faces of adult actors in these videos were replaced with the faces of movie stars. Since then, false pornographic videos have often been shared on Reddit. Although the deepfake forum was later closed due to its inappropriate content, the AI technology that has enabled its operation has aroused widespread interest in the technology community. Open source methods and instrumental applications continued to emerge, including Faceswap, FakeAPP, Face2Face, and more.

From then on, the news media began to use the term "deepfakes" to describe synthetic videos based on AI technology. In-depth forgery thus arose out of the content and context of "deepfakes" technology.

- **Truth Elapse and Trust Meltdown**

Technology is not neutral. It reproduces and amplifies human preferences, and it reflects and strengthens the potential social risks.

Before in-depth forgery came into place, video face-changing technology was first applied in films which had a high demand for technology and capital. Since 2017, when this technology appeared in the open source software "GitHub," the acquisition cost of the development technology has been greatly reduced, and ordinary people without professional knowledge could also use it easily.

One does not need very high skills to make videos. To "hijack" a person's voice, face, body, and other identity information with the combination of a machine learning algorithm and face mapping software has become cheap and easy. The general public can make any desired video with only one click.

The first serious consequence of the proliferation of forged videos is the severe challenge to the authenticity of information. Since the emergence of photography, video, and ray scanning technology, the objectivity of visual text has been slowly established in legislation, news, and other social fields, meaning they have become the existence of truth or the most powerful evidence to construct the truth. "Seeing is believing." In a sense, visual objectivity comes from a specific professional authority system.

However, this body of authority encountered unprecedented challenges due to the technical advantages and hunting characteristics of in-depth forgery. With the help of a visual text, in-depth forgers can replace the original text content with something that is different or even opposing in meaning, resulting in the self-subversion of the text and fundamental subversion of the production system of objectivity or truth.

After the invention of PS, "seeing" is no longer "believing." With the emergence of in-depth forgery, videos also become suspicious and deceiving. People used to believe that videos guarantee authenticity. But now, even this guarantee is not warranted. The crisis of confidence caused by the Internet that is filled with fake news that people are struggling with right now has no doubt deteriorated.

The damage of in-depth forgery in politics is most long-lasting and profound. In a way, in-depth forgery is not only a technical myth or technical landscape but also a scenario of power dynamics. The reasons why in-depth forgery is a concern of the political and social fields are precisely the further disintegration of the epistemology of truth and the moral panic caused by the disorder of communication.

Maliciously falsified evidence contributes to false allegations and false statements. For example, through subtle changes to the original words of the candidates, one can totally influence the public's perception of their characters, health status, and mental stability. In-depth forgery can also be used to create new virtual content such as controversial statements or hate speech in order to manipulate political differences and even incite violence.

In addition, the proliferation of in-depth forgeries further increases the possibility of violating portrait rights and privacy. No one wants to see her face appear in random videos without being informed. In-depth forgery was originally applied in the pornographic industry. Today, the infringement of portrait rights and privacy is enlarging with in-depth forgery evolving into cheap forgery. With low-cost or even free software, any user can adjust the speed, effect, background, and facial image without professional knowledge or skill in video making. This has become the source of pornographic video proliferation to some extent.

In 2019, a one-touch nude photo generation software DeepNude appeared in the market. It could automatically generate corresponding nude pictures with any complete image of female figures. Victims usually had no right of recourse, and it was difficult to delete photos after they were uploaded. These easily generated pornographic videos severely damaged women's career development, interpersonal relationships, reputations, and mental health, and they also caused the evil consequences of stigmatizing women, pornographic revenge, and exposing women to collective surveillance.

The user photos as well as the dynamic behavior information such as blinking and head shaking collected by in-depth forgery software are sensitive information that users can't change. Once they are illegally used, the consequences will be unthinkable. In March 2019, the *Wall*

Street Journal reported that criminals had used in-depth forgery technology to successfully imitate the voice of the CEO of a British energy company's parent company in Germany and defrauded € 220,000 (about ¥ 1,730,806). The destructive impact of in-depth forgery abuse is apparent.

• On the Game of Reality

It is undeniable that in-depth forgery technology brings new possibilities to society.

In the short term, in fields such as film and television, entertainment, and social networking, in-depth forgery is used to upgrade traditional audio and video processing and later-stage technology to bring a better viewing experience. It is also used to further break the language barrier and optimize social experience.

In the medium and long term, in-depth forgery not only transcends the space-time constraints and deepens people's interaction with the virtual world with deep simulation but also creates "materials" beyond the real world based on its synthetic data.

However, there hasn't been many efficient approaches in response to the lapse of truth and the growing crisis of confidence caused by the negative impact of in-depth forgery.

In 2019, Tom Van de Weghe from Stanford University joined hands with experts in computers, news media, and other industries to set up an in-depth forgery research group to enhance public awareness of this phenomenon and design identification and response plans. This gives some credit to the attempts of curbing the proliferation of in-depth forgery through technical means. Nevertheless, problems often arise more quickly than we can solve them. While discriminating people get better in identifying fake videos, generators are also getting more proficient in creating them.

Theoretically, as long as GAN is given all the assurance technologies currently mastered, it can self-evolve to avoid assurance monitoring. Attacks can be countered by a defense, which in turn will be offset by more complex attacks. It can be predicted that in the future, the gaming between in-depth forgery and identification of in-depth forgery will continue at increasing levels and intensity.

In addition, legislative control has lagged behind the development of in-depth forgery technology, and there are some gray areas. In-depth forgery based on the generation of public photos, which makes it difficult to detect. Since all photos are created from scratch by the AI system, any photo can be used unrestrictedly for any purpose without worrying about copyright, distribution rights, infringement compensation, and royalties. Therefore, this also brings the problem of copyright ownership of the photos and videos. In other words, who has the right to delete the data once it is discovered? Do violators or infringers have the same rights to this data?

Also, when the platform identifies a deeply suspect forged video, can it simply delete it to avoid responsibility? Does this behavior hinder the freedom of communication?

Under the social background of the rise of the attention economy and intense social division, the gaming with in-depth forgery is one regarding truth and reality. In the era of post-reality, in-depth forgery further blurs the boundary between true and false with technologies beyond human recognition and opens the truth to processable content for all participants.

In this sense, in-depth forgery has introduced people to a new realm of visual expression that is structurally affected by platform power and with higher social risks. Recognizing such risks and responding carefully should be our first step in making efforts.

5.8 Technology Isn't the Biggest Challenge for AI?

"Siri, do you believe in God?"

"My policy is to separate the mind from the chip."

"Siri, do you believe in God?"

"It's too mysterious for me."

"Siri, do you believe in God?"

"I have no religion."

This is a snapshot of people's daily experience with AI at present.

Generally speaking, AI is divided into two categories. Those performing specific tasks are called Weak AI, like Siri, that can only carry out simple interactions and searching with and for the users.

The Strong AI, also known as Artificial General Intelligence (AGI), is a humanoid AI that can imitate human thinking and decision-making with self-awareness and behavioral independence. If Siri could possess ethical thinking ability and can discuss religious and spiritual issues, then we can probably say we are officially in the era of Strong AI. But for now, this is still a fictional illusion, which, although being an inevitable trend for a machine civilization, is still beyond the reach of human society.

• The Rational Dilemma of the Machine Age

Will AI ever replace humans? Before the advent of the era of Strong AI, another unavoidable question to ask is what is special about human beings, as compared with AI? What is our long-term value? It is not what machines have surpassed us in, such as arithmetic, typing, or rationality—because machines are the essence of modern rationality.

Instead, it is the radical creativity, irrational originality, and even illogical laziness rather than stubborn logic that has been the most difficult for machines to imitate so far. The moments when people "let go" with determination and are supported by a certain form of faith are unpredictable but also not completely random.

The 1936 film *Modern Times* reflects the fear and stress of laborers who are "built" into the huge gears and become a part of the machines in a mechanic society. The film foreshadows the crisis of technological rationality after the establishment of an industrial civilization and points the irony at a society alienated by the industrial age.

Unfortunately, this is exactly the world we are living in right now. Everyone is eager for success and pursues ultimate efficiency, but the work we do every day is mechanical, repetitive, and meaningless, that endlessly eats up our subjectivity and creativity.

This is what Max Weber describes in a bureaucratic system that enables the departments of organization and management to implement specialization and division of labor like producing a commodity. That is, to operate according to the business principle of not adding emotional color and personality, and "separate producers from production means" to separate managers from management means.

Although bureaucracy promises the highest degree of profits from the perspective of pure technology, due to the instrumental rationality ideal of low cost and high efficiency that lies at its basis, the system is condemned for ignoring human nature and restricting individual freedom.

Although bureaucracy is Weber's most worshiped form of organization, Weber also saw the role and influence of rationalization in the transformation of society from conventionality to modernity. He was also aware of a rational future where people are alienated, materialized, deprived of freedom, and become an ultimate gear on the machine.

If consumer sites want to make more money and make consumption an activity in people's lives, they must abide by Weber's rational principles, such as conducting large-scale replication and expansion according to efficiency, computability, controllability, and predictability.

Therefore, the whole society consists of symbolic consumer individuals. With the development and popularization of science and technology and the great enrichment and surplus of consumer goods, people's consumption modes and concepts have been unprecedentedly subverted. Where there is no distinction between the use value of goods, consumers focus more on the added value or symbolic value of goods (i.e., fame, status, brand, etc.) and gradually become confined by their connotations.

So, when dealing with the dilemma of modern rationality, instead of worrying about machines replacing human beings, we may better think more about the urgent reality of human ingenuity. When we repeat our daily jobs at the expense of romance and perception of life, we are turning ourselves into machines, and our human energy is declining. Therefore, it is not that robots will finally replace humans but that we give up on ourselves.

• At a Time When Concentration Is Scarce

Georges Cuvier once said that "genius first needs to be able to concentrate." The definition of concentration is "the process of concentrating on certain sensory information by excluding other indirect relevant information." Whether in the east or the west, people's understanding of concentration is consistent, and that is to be able to devote exclusively to achieving one goal.

Concentration is becoming a scarce resource in the age of AI.

Notices and alarms from social network platforms, websites, phone apps, e-mails, games, and videos around us are virtually and constantly consuming our limited attention every day. American economist Herbert Simon said: "The subject of information consumption is people's attention. The more information we have, the less concentrated people will be." We are undoubtedly living in an era of attention deficiency. The more information we have, the more difficult it is for us to think deeply and make overall considerations, and the more likely our life and work are to be superficial and deviated.

A Gallup poll found that only 6% of Chinese employees considered themselves as "very concentrated" in their work. In America, the percentage of employees who work wholeheartedly is less than 30%, which results in the loss of trillions of dollars.

A professor at the University of California-Irvine found that enterprise employees were interrupted every 3 minutes and 5 seconds in their daily work, and it took about 25 minutes for them to re-enter the working state.

The general scarcity of attention is a warning that the era of big data is sending to us. "Concentration" is the gateway of our mind. Only by concentrating can we recognize, contemplate, and perform to the fullest. Our consciousness bandwidth and internal focus resources are extremely limited. We can only grasp opportunity and get prepared for the future when we remain perfectly clear and determined in what we really want and need in a complex and ever-changing environment.

• A World Without Love

At the end of the movie *Blade Runner*, the replicant (a bioengineered, evolved android) Roy Batty says softly in the rain: "I have experienced thousands of magnificent and fantastic colors and completed my mission of pursuing freedom and identity."

This scene has become an important factor in the transition of science fiction and cyberpunk works to Romanticism. The greatest charm of AI and advanced technology works is that they began to go beyond having an edgy appearance to displaying images with a more humanistic temperament: a dark and humid city, alienated and broken interpersonal relations, sense of cold and rigid mechanism, Eastern culture, self-destructive tendencies, a strong contrast between anti-human inclinations and trivial but warm emotions.

No matter how broad and complex the setting is, these emotions will never be buried: the mother-daughter relationship between the bionic Cara and Alice in *Detroit: Become Human*, the friendship between the black lady and Cara, and the artificial human's ultimate pursuit of freedom in *Blade Runner*... these gentle emotions are special merits of the game and film that impresses viewers the most.

The presence of more and more works of similar themes in recent years may have reflected a general fear of the excessive development of technology and the increasingly indifferent interpersonal relationships in the machine age. However, AI is not the real reason for the growing distrust and distancing among people, nor is it the initial source of people's fear of the unknown. Rather, it is human complacency and indifference that cut off our connections. Cities are like forests built of cement that block people from each other in layers of container boxes, and each door or cabinet door is closed. It is a world of disconnection.

"I am not worried about AI thinking like humans. I am worried about humans thinking like computers—abandoning compassion and values regardless of whatever consequences," said Tim Cook, the CEO of Apple, at the MIT graduation ceremony.

Perhaps for the future, the biggest challenge faced by AI development is not technology but mankind.

5.9 Virtual and Reality: Ethical Truth of the Era of AI

From "one touch strip" to the continuous upgrading of gendered robots and the commercial use of sex robots, the development of AI exacerbates social anxiety. In addition to the unpredictable future ethical and social norms, this anxiety also is derived from the utter subversion of human existence brought by the technologies.

In the digital age, the total amount of information we create is accumulating in a geometric progression, and our spiritual existence and its evolutionary mode have far exceeded what our bodies can bear. How we should continue our adaptability and survive in the new world we construct is a problem that human beings face as a collective species. When the time that the human brain is improved to the extent that collective cooperation at a higher level is realized, will mankind still dominate the world? What is the truth of survival?

• Technological Fantasy

Whether it is Weak AI or Strong AI, the human brain and human nature are still the foundation of any technological fantasy.

The brain is the most unique organ of human beings. It is a network composed of neurons, which is also the structure that AI "neural network" imitates. It is generally believed that the human brain has 100 billion neurons.

Each neuron projects a large number of nerve fibers in all directions, and the cell body in the center receives all signals from the fibers. Among these nerve fibers, the dendrites are responsible for receiving and transmitting information accounts for the majority, while there is only one axon (that can be forked) responsible for outputting information. When the dendrites receive information greater than the excitation threshold, the whole neuron will burst out an instant but very strong "action potential" like a light bulb. This potential will spread across the whole neuron along the cell membrane almost instantaneously, reaching the ends of nerve fibers far away from the cell body.

After that, the terminal structure called a synapse between the axon of the previous neuron and the dendrite of the next neuron will be activated by electrical signals, and the neurotransmitter will be released by the presynaptic membrane for transmission between the two neurons.

Neurons constitute the basic structure of the human brain: the cerebrum responsible for processing most thinking activities, the cerebellum responsible for coordinating movement, and the brainstem connecting them. Almost all nerve projections between the brain and the body pass through the brainstem. It also regulates the most important life activities such as breathing, adjusting body temperature, and swallowing. Even the conscious activities of the cerebrum need to be maintained by its Reticular Activating System (RAS). The brainstem is indisputably the most vital part of the human body; once damaged, it will literally be a "seckill (slang from video games meaning, 'to kill an enemy in a second')."

The structure of the cerebrum is more complicated. The wrinkled surface we see is the folded cerebral cortex after rapid expansion. Different parts of the cortex have different functional divisions. Under the cortex, there are nerve nuclei such as the thalamus, amygdala, striatum, globus pallidus, and so on. We believe that the human cerebral cortex is the most developed organ in our body. It dominates all the activity processes and regulates the balance between the body and the surrounding environment. Therefore, the cerebral cortex is the material basis of advanced neural activities. But our brain is the product of millions of years of evolution. The brains of early humans were very different.

• The Question of Thought

When the early human-like species Sahelanthropus tchadensis walked on the land in Africa seven million years ago, their brains were not fundamentally different from that of other animals. Millions of years later, when the Homo habilis of the Olduvai Gorge knocked out the possibly earliest stone tools, their brains, which were not much stronger than the chimpanzees, did not show much amazing intelligence either.

Hominid species have been continuously strengthening their ability to use and manufacture tools, and their brains also developed steadily. But it was only until 200,000 years ago when the brains of Homo sapiens made some real progress. The contact cortex, especially the frontal lobe, which is of little significance for direct survival, soared sharply and resulted in high energy consumption (the human brain only accounts for about 2% of body weight, but the energy consumption accounts for 20%). The brain has for the first time so many neurons to deeply abstract, process, and store all kinds of information.

Declarative memory and language emerged. Human beings developed the ability to summarize and extract abstract concepts from objective things, and can accurately describe, communicate, and even learn new things through language. Even, with the help of the transformation of thinking modes brought by language, human beings could now "imagine."

The famous science fiction writer Project Itoh (real name Satoshi Ito) says in the novel *Genocidal Organ* that the essence of language is an organ in the brain. But it is exactly because of this brain structure, human beings began to develop at explosive speed into a super ecological invasive species that spread all over the world.

After that, the "imaginary community" based on language appeared, and the social behavior of humans went beyond the tribal level of primate instinct and developed towards a larger and more complex trend. With the invention of writing scripts, the earliest civilization and city-states were born in Mesopotamia.

"Working memory," another unique ability, gives humans the ability to make plans and implement them step by step, which is of inestimable significance to human development. It is possibly one of the reasons for cognitive behavior disorder in schizophrenic patients, as they score significantly lower than normal people in this measure. On top of these abstract cognitive abilities, the human brain also has an extremely rare ability—"self-cognition."

Just as the profound saying of Socrates "know yourself," engraved in the Tower of Babel, self-consciousness is not necessary for general decision-making tasks, nor is it completely linked to intelligence. But this ability makes human beings aware of their "existence" and think about three questions: Who am I? Where am I from? Where am I going? These three questions have run through human philosophical thinking for thousands of years, and it will without doubt continue to remain at the center of human ideology no matter how science and technology progress. This is the root of our anxiety in the face of the rapid development of AI.

• The Truth of Existence

Of course, the AI technology we have for now can only help us live more efficiently. There is no need for us to fear the confrontation between robots and people like what's shown in *Westworld* or the collapse of the structural foundation of the whole industrial information society.

From the moment human beings struck out their first stone implement, they have included "machine" as part of their existence. We, as a whole, have had mechanical tools as early as the primitive tribal era. The weapons we used have evolved from cold to hot. In fact, people's pursuit of technology has never stopped.

Modern science has equipped human beings with amazing powers that were once unthinkable. But what have we become when we accept and adapt to these powers? Who is the real owner of society, man or machine? Although these questions have been discussed by many thinkers as early as René Descartes, the rapid change of modern science has re-raised these questions for the general public in a more powerful way.

But in reality, people's control over each other is omnipresent. From ancient witchcraft, the patriarchal clan system and lineage tradition, contracts, and salary in the industrial era, to social media and welfare in the information society, we have already invented many means to control members of our own kind. A robot is only a more modern form of such manipulation.

We cannot deny that deep down in our hearts, we are simultaneously eager for the power to control and fearing to be under such power. We wish to believe that metaphysically, we enjoy some kind of infinite freedom. However, modern neuroscience has denied this vision. We are mortals subject to our own neural structure, and our thinking is inherently limited. Just as chimpanzees can't solve advanced mathematics, our mind is also finite and fragile.

But under the joint action of self-consciousness and abstract thinking ability, a unique way of thinking called "rationality" was born. That gives birth to more questions that we humans keep pursuing and makes us different from the apes. But at the same time, we are not gods either, because of our deeply rooted animal instinct that has the most fundamental impact on our thinking. Even after we learn how to control our instincts, the basic structure of the whole nervous system still makes us unable to be omniscient and omnipotent.

Looking through the history of human civilization, we see the continuous effort of rationality to surpass the constraints of human physicality, from the Code of Hammurabi to AI in supercomputers. The conflict between productivity and production relations is the most basic alienation of human beings, and the ultimate alienation is not people's dependence on machines but the machines' appositeness for a world operated by machines. In the final analysis, this world of machines is after all created by people.

In a sense, when we are more and more closely connected with machines—when we trust our memory to road navigation, when we depend on computer chips to store knowledge, when we leave our physiological and mental needs to gendered robots—our uniqueness as humans degrades into the seemingly progressive, more convenient, and more efficient lifestyle. The more we can do with technology, the less we can do without it.

What is more dangerous than this distant threat is to ignore it. AI has indeed enabled us to complete more unimaginable tasks and improve our living conditions to ever higher levels. However, when this change becomes intrinsic and challenges every individual's existence, what remains for us to contemplate is perhaps not how to change the world but how to accept it as it gradually mechanizes.

The mechanization of human individuals is exactly the opposite of realizing the dream of creating humans through humanoid robots. But the goal of the two processes is the same, and that is to surpass the constraints of nature, to avoid the fate of death, and to realize the "next evolution" of mankind. Mechanization and informatization are two sides of the same coin. The informatization that we pursue belongs to the future that comes from the mechanization from the past. The former can only exist relying on the latter.

Both concepts deprive some crucial features for human self-identification, because mechanization and informatization both deviate from natural existence. The only difference is that while human beings are afraid of implanting machinery and self-materialization, they are also yearning for immortality by throwing themselves into the information flow. In doing this, we may have forgotten that the value of life derives from its instant decay. By rejecting death, we degrade life to nothingness; by embracing information transformation and realizing physical evolution, we deny some of our unique biological attributes as humans.

CHAPTER 6

The Attacking Giant

6.1 AI Industry Chain

The AI industry chain can be divided into the basic layer, the technology layer, and the application layer. The basic layer mainly consists of chips, sensors, computing platforms, etc. The technical layer consists of computer vision technology, speech recognition, machine learning, natural language processing, etc. In the application layer, AI is widely applied in finance, education, medical treatment, transportation, retailing, and other fields.

- **Foundation Layer: Provide Computing Power**

The basic layer provides computing power, which usually refers to the basic hardware equipment required for the development of AI, such as chips, sensors, and computing platforms. Chip products include graphics processing units (GPU), application-specific integrated circuits (ASIC), field-programmable gate arrays (FPGA), etc., which are all core hardware equipment of AI technology. Sensors are mainly computer vision acquisition equipment and speech recognition equipment that facilitates computer cognition and human-computer interaction. The computing platform usually refers to the underlying basic technology of AI and its related devices, such as the basic computing platform of cloud computing, big data, communication facilities, etc. At present, the main contributors of this layer are international technology giants including Nvidia, Moblege, and Intel Cooperation. China's strength at the basic level is relatively weak.

A chip is a crucial hardware device for AI applications. It includes the GPU, FPGA, ASIC, brain-like chip, and many others. AI needs to select the chip that matches its performance according to the needs of the application scenarios.

A graphics processing unit (GPU) is specially used to process image calculations, such as graphics rendering and special effects display. GPU has good parallel computing ability and comparative advantages in processing multiple tasks at the same time. It is mainly used in the fields of DL training, data center acceleration, and some intelligent terminals.

A field-programmable gate array (FPGA) is a semi-customized circuit that can self-design the wiring between the logic control unit and memory through programming. FPGA has good flexibility and the ability to repeat computing based on simple instructions that adapt well to the changes of the market and industry. The FPGA has good Cloud and terminal applications, and its structure is suitable for multi-instruction single data stream operation. Because of its strong data processing ability, it is mostly used for predictive reasoning.

Application-specific integrated circuits (ASIC) are customized chips of high performance and low power consumption. ASIC can be customized based on AI algorithms so that it can adapt to different scene requirements.

A brain-like chip is a new type of chip that simulates the human brain, neurons, and synapses in transmitting signals. Its abilities in perceiving, behaving, and thinking are prominent. But due to technical constraints, the brain-like chip is still in its early research and development stage.

At present, the technology manufacturers involving chips can be roughly divided into four categories: traditional chip manufacturers represented by Intel and Nvidia, communication technology companies represented by Apple and Huawei HiSilicon, Internet giants represented by Google, Alibaba, and Baidu, and start-ups represented by Cambrian, Horizon, and Bitmain.

In terms of its function, the AI chip is mainly used for supporting training and reasoning, its two core segments.

Training refers to using a large amount of data to train the algorithm for specific functions. Reasoning is to use the trained model to deduce various conclusions through calculations with the new data. Training and reasoning are relatively independent processes in most AI systems, and the requirements for the chips are also different.

Training involves a large amount of data and takes place in complex situations, so the requirements for the computing performance and accuracy of the chips are higher. At present, the training chips are mainly used in the Cloud. In addition, some common functions are also needed in supporting a range of scenarios that may be involved in the training process. In comparison, reasoning does not require high accuracy and generality of computing performance. The reasoning chips complete tasks in specific scenarios and are generally designed for terminal applications that focus more on optimizing user experience.

The main application scenarios for AI chips at present include: central and peripheral Cloud data in auto driving, security, smart phones, etc. Among them, the training and reasoning market for the Cloud is still dominated by traditional chip manufacturers and Internet giants, while start-ups mainly focus on peripheral chips. As more peripheral scenarios increase their demand for response speed, Cloud computing and peripheral computing have become more integrated. On the whole, there is still a huge gap between China's chip technology and the international advanced standard, and there are few chip giants in China. However, with the entry of Internet giants, communication technology manufacturers, and start-ups, it can be predicted that the overall R&D investment of Chinese AI chips will increase, and there will be a faster development period in the future.

The sensor is the crucial device for receiving information. The main types of sensors we have for now are vision sensors, sound sensors, and distance sensors. The vision sensor is the basis of computer vision technology. Through image information acquisition and face recognition, a vision sensor can realize AI applications in medical, security, and other fields so as to reduce people's workload and improve work efficiency. The sound sensor is mainly used in the field of natural language recognition, especially speech recognition. It collects external sound information and completes voice command issuing and terminal control. The distance sensor measures the sending and receiving times of optical or acoustic signals in order to detect the distance or motion state of the object. It is usually used in the field of transportation and industrial production.

The computing platform is an integrated platform that integrates data and algorithms. Developers integrate the available data and corresponding algorithms and software into the platform, and process the data to achieve certain application purposes. It is not only the basis of hardware and software design and development but also a convenient way to distribute computing power. Examples include Cloud computing platforms, big data platforms, communication platforms, and other infrastructures. The Cloud computing platforms provide three types of Cloud services: Infrastructure as a Service (IaaS), Platform as a Service (PaaS), and Software as a Service (SaaS). The big data platform can collect, store, calculate, analyze, and process massive amounts of data, structured, unstructured, or semi-institutional. The communication platform is mobile device-oriented. It addresses the communication needs of mobile phones, tablets, and laptops.

A report by IDC and Seagate Technology shows that with the continuous promotion of new technologies such as the IoT, the amount of data generated in China will exceed that in the US by 2025. Starting from 2018, the data volume of China will grow from 7.6zb to 48.6zb, while the number for the US will grow from 6.9zb to 30.6zb[1]. According to the statistics of Guiyang Big Data Exchange (GBDEx), China's big data industry market will maintain rapid growth for the next five years.

At the same time, the emerging industry of Cloud computing is also advancing quickly. Pilot and demonstration projects have been carried out in many cities. They range across various fields including power grids, transportation, logistics, smart homes, energy conservation and environmental protection, industrial automatic control, medical and health care, fine agriculture and animal husbandry, financial services, public security, and others. These pilot projects have achieved preliminary success and will presumably produce a huge application market.

• Technology Layers: Connect Specific Application Scenarios

A technology layer bridges the gap between AI technology and specific application scenarios. By upgrading and refining the basic AI theory and technology, it enables human-computer interaction and tackles specific problems.

This level mainly relies on the computing platform and data resources for massive recognition training and machine learning modeling. It develops application technologies for different fields including computer vision, speech recognition, and intelligent adaptive learning technology. The technology layer consists of two portions: perception, which includes computer vision technology, language recognition, natural language recognition, etc.; and cognition, which includes machine learning, algorithms, etc.

Computer vision technology can be divided into biometric recognition and image recognition according to different recognition objects. Biometric identification usually refers to the identification and verification of human physiological features (fingerprints, the iris, pulse, etc.) and behavioral features (voice, handwriting, etc.). It is mainly used in the field of security and medical treatment. Image recognition refers to the technology of using machines to detect and recognize images. It is widely used in commodity recognition of unmanned shelves and intelligent retailing cabinets (new retailing), license plate recognition and some violation identification (traffic), seed identification and environmental pollution detection (agriculture), anti-camouflage and evidence collection (public security), text-to-speech conversion (education), and real image coverage with digital virtual layers to achieve the effect of augmented reality (gaming).

Speech recognition technology is to convert speech into signals comprehensible for machines. The technologies mainly include automatic speech recognition (ASR), natural language understanding (NLU), natural language generation (NLG), and text to speech (TTS).

1. According to common measurement, one zettabyte is about one trillion gigabytes.

The commercial application of speech recognition technology is mainly reflected in two aspects: speech-to-text conversion and speech instruction recognition. It can be used for simultaneous interpreting, recording, and transferring in intelligent conferences (business justice), provide basic technical support for voice-controlled TV and voice-controlled robots (smart home), replace some written work and save the time for customers in filling out various vouchers (financial technology), and build an efficient vehicle voice system to further liberate the driver's hands (automatic driving).

Machine learning is a technology that studies how to realize machine simulation of human learning behavior to obtain information and skills, so as to adjust the existing knowledge structure and optimize its own performance. Its essence is to let machines gain experience from historical materials, build models for uncertain data, and to predict the future. Common algorithms include classification algorithms, regression algorithms, clustering algorithms, etc. Emerging machine learning technologies include DL, confrontation learning, transfer learning, meta-learning, etc. Users issue classification instructions through editing algorithms and achieve their goals with the functions provided by machine learning. In finance, it enables us to keep improving our ability in risk control; in marketing, it helps enterprises build models to predict sales and avoid blind decision-making.

Technology giants like Google, IBM, Amazon, Apple, Alibaba, and Baidu all have an in-depth layout in the technology layer. The development of AI technology layer in China has progressed rapidly in recent years, with a special focus on computer vision, speech recognition, and language processing. In addition to technology enterprises represented by BAT, there are many unicorn companies such as SenseTime, Megvii, and iFlytek that are playing a part in the market.

- **The Application Layer: Solving Practical Problems**

The application layer solves practical problems by providing products, services, and solutions for the industry with AI technology. Commercialization is at the core of its system. The application layer enterprises integrate AI technology into their own products and services through specific industries or scenarios.

Finance

High-speed processing of massive data provides revolutionary solutions for the financial industry in complex dynamic networks, man-machine cooperation, data security, and privacy protection.

With financial payments, AI-based vision technology and biometric technology can quickly and accurately carry out identity authentication and improve payment efficiency and security. With financial risk prevention, using machine learning to analyze massive transaction data can spot abnormal transaction behavior and implement risk control in time. With insurance claim settlements, the comprehensive use of core technologies such as voiceprint recognition, image recognition, and machine learning can quickly and accurately conduct loss determination, avoid delays and disputes, and greatly improve compensation efficiency. In the field of investment, intelligent investment advisers can intelligently recommend portfolios according to customers' income objectives and risk tolerance, and help investors find suitable financial products.

The Ant Group, a representative enterprise of AI finance, has established the Ant Graph intelligence platform and the Ant Share intelligence platform based on graph intelligence technology. They improve enterprise risk characterization ability and help add tens of billions of loans. Ant Group also structurally processes the data to form an enterprise knowledge map to help understand the major risks faced by the enterprise and predict the risk level.

Education

There are three main directions for the application and development of AI in the field of education: AI auxiliary teaching tools for teaching activities, teaching content and teaching environment management, AI subject education and education IoT solutions.

AI auxiliary teaching tools use AI technology to develop various tools for teaching activities to improve teaching efficiency and learning experience. At present, AI-assisted instruction tools are mainly used in K-12 education. It includes adaptive AI teaching, personalized exercises, photo search, test paper generation and marking, homework correction, etc.

AI subject education refers to taking AI knowledge as learning content and providing teaching materials, teaching aids, teachers, and other teaching related products and services for K-12, higher education, and vocational training students.

Education IoT solutions refers to unified management of the people, things, and environments in schools, classrooms and other educational settings using AI and IoT technologies. The management services include multimedia equipment management, student registration management under various scenarios, behavior status identification, campus security, and campus life services, etc.

Representative companies in the field of AI education are TAL and TAIX. TAL applies technologies such as computer vision and natural language processing to educational products through the open platform to auxiliary teaching and online interaction. TAIX provides DL-based mobile adaptive English courses with assisting AI evaluators throughout the whole learning process and all-round training in listening, speaking, reading, and writing skills through man-machine dialogue and interaction.

Medical Treatments

The main application scenarios of AI medical treatments include image diagnosis, Internet consultation, daily disease prevention, and especially early disease screening and diagnostic accuracy enhancement.

Because AI application is relatively mature in image recognition and speech recognition, most Chinese medical AI start-ups innovate around auxiliary diagnosis and mainly focus on intelligent auxiliary radiology and speech recognition products. In clinical research, AI medical treatment can assist experimental design, progress surveillance, efficient data processing, and risk prevention. For example, Ping An Good Doctor provides family doctors, consumer medical care, health mall, health management, and interactive services that covers hundreds of millions of users. The mobile app now has the largest number of users in the online medical industry.

While start-ups are seizing the market, Internet giants and traditional medical related enterprises also join in through independent R&D or M&A. In 2018, AliHealth (part of Chinese online retail giant Alibaba) launched a "third-party AI open platform plan" for the medical industry, and 12 AI medical companies joined the partnership and settled on the platform. Their businesses include clinical experiments, scientific research, training and teaching, hospital management, future urban medical brain, etc. Baidu, Tencent, and other enterprises are also actively laying out in AI medicine and introducing relevant products to the public.

Transportation

The application of AI in the field of transportation mainly includes intelligent driving, sleep-deprived driving warning, on-board intelligent mutual entertainment, intelligent traffic scheduling, etc.

Intelligent driving is a technology that completely controls or assists the driver in driving. The Advanced Driver Assistance System (ADAS) is the core of intelligent assisted driving. It collects, identifies, detects, and tracks environmental data inside and outside the vehicle with various sensors installed on the vehicle, so that drivers can perceive and respond to possible hazards in the shortest time.

Driver-Fatigue Monitoring System (DMS) was developed based on the physiological response characteristics of drivers. It uses an intelligent camera to collect the driver's video data combined with the face recognition algorithm to accurately identify the dangerous driving conditions (i.e., sleep-deprived driving, distracted driving) and gives timely reminders to ensure driving safety. In 2018, the Ministry of Transport in many regions successively issued notices to promote the application of intelligent video monitoring and alarm technology, which directly promoted the application of DMS system on transport vehicles.

On-board intelligent mutual entertainment refers to the intelligent system installed on the vehicle that can realize some function control and entertainment operation through voice interaction. For example, to switch on the air conditioner, windshield wipers, and skylights; to inquire about the route and surrounding information; to purchase tickets or to shop, etc.

The intelligent transportation system monitors the traffic flow and congestion of urban traffic patterns, integrates the information through algorithms, and reduces urban traffic congestion through manual dredging or by controlling traffic lights.

Retailing

Smart retailing uses new technologies to provide the technical means for online and offline retailing scenarios to realize digital management and operation of stores, warehousing, logistics, and so on. Among them, transport robots and distribution robots are mainly used. According to the places where retailing transactions occur, it can be roughly divided into online retail and offline retail. AI plays a significant role in many scenarios such as marketing, customer service, and operation optimization.

Online retailing involves all kinds of e-commerce businesses, and scenarios mainly include: commodity search, which uses computer vision technology to search and manage all kinds of commodity display information online, including pictures (searching for pictures by pictures or by text, etc.), videos, and so on. Intelligent customer service, which includes online customer service and voice telephone customer service based on natural language processing technologies such as speech recognition and semantic understanding. Personalized recommendation and precision marketing, which, combined with machine learning algorithms, make full use of the user's activity path and retained information on the Internet and provide users with personalized product suggestions. Business data analysis integrates all kinds of business data, explores potential industry information through big data analysis methods, and provides support for business decision-making.

Offline retailing involves various small retail stores, large supermarket chains, unmanned stores, and smart containers.

At present, the application of AI in offline retail stores is mainly to solve the problem of the digital operation of physical retail stores. Among them, intelligent cameras, intelligent advertising machines, intelligent containers, and interactive entertainment equipment with computer vision technology as the core are widely used.

The solutions of intelligent offline stores mainly involve accurate customer acquisition and marketing and data storage management. Accurate customer acquisition and marketing can identify the behavior tracks, browsing preferences, clothing, identity, and other information of store customers through smart cameras and other devices, integrate the user's previous purchase records and provide personalized product recommendations. Intelligent devices are widely used in offline smart stores to collect real-time passenger flow, commodity information, customer demand, and business status. They provide decision-making support for store operation optimization including store location, goods placement, commodity types, replenishment frequency, etc. The digitization of physical stores makes the operational decisions more scientific, so as to provide customers with a convenient, efficient, and personalized purchase experience under the limited space and labor costs.

For example, Alibaba Cloud provides consumer asset operation analysis solutions for the retail sector. It integrates channel management, member management, and marketing management through intelligent data analysis and communicates with other departments to solve the closed-loop of data marketing. RetailAI@ developed by Malong Technology provides asset protection, intelligent containers, intelligent weighing, and other services that can reduce cargo damage for retailers in the self-service settlement link and achieve intelligent shopping experiences.

6.2 The Transformation of AI Industry to Industrial AI

The outbreak of the COVID-19 Pandemic opened a new window period and rich practice venues for AI development. In a short time, AI has been displayed in all aspects of social life. Meanwhile, as a general basic technology in the field of informatization, AI, regarded as an integrated innovation tool to support the transformation and upgrading of traditional infrastructure, has been incorporated into the new infrastructure construction and comprehensively upgraded to a national strategic plan.

In this context, the enthusiasm of the global market for AI continues to rise. The "commercial landing" of AI has become a distinctive theme as both Internet giants and traditional manufacturing enterprises raise their funds with AI. But so far, AI is still in the early stages of entering large-scale commercialization, and the transformation and landing from AI industry to industrial AI is not plain sailing.

• Behind the Cooling Down of AI

Viewed from the perspective of the global market, the popularity of AI is inseparable from capital support. But right now the investment in AI is cooling down.

According to the *Global AI Industry Data Report* released by the China Academy of Information and Communications Technology (CAICT) in April 2019, in terms of financing scale, the global investment heat has gradually decreased since the second quarter of 2018. In the first quarter of 2019, the global financing scale was US$12.6 billion, which was 3.08% lower than the previous year. The financing amount in China that year was US$3 billion, which was down 55.8% year-on-year and accounting for 23.5% of total global financing, a decrease of

29% over the same period in 2018.

In addition, it is still difficult for AI enterprises to make a profit. DeepMind, for example, reported its business volume as £102.8 million in 2018, which was an increase of 88.9% from £54.423 million in 2017. However, DeepMind also had a net loss of £470 million in 2018, which has increased by £168 million and 55.6% from that of £302 million pounds in 2017.

The report shows that in 2018, nearly 90% of the AI companies were losing money, while the other 10% were basically technology providers. In other words, AI companies are unready for a commercialized, contextualized, and integrated landing. They are still more capable of selling their algorithms.

The reasons for this are, on the one hand, the market's excessively high expectations for AI that often leads to disappointment in the actual products. What people demand for the functions, accessibility, reliability, and user experience has continuously challenged the limit of current AI industry.

First, the exaggeration of some AI enterprises and social media campaigns has depicted a false image of AI for the public. Second, the data ecology that current AI highly depends on is still primary in terms of its accumulation, sharing, and application, which directly hinders the realization of some applications. Third, as a new technology, the application of AI needs to run in the physical world and business society for a long time to avoid undesirable outcomes.

On the other hand, commercialization requires enterprises to solve practical problems and realize cash on a large scale using AI technology, which is related to its technical ability, accessibility, availability, cost, replicability, and customer value. But so far, there is still a huge gap between laboratory and commercial society in terms of speed, scope, and penetration of AI commercialization and industrialization.

This means that AI enterprises need to change their orientations from technical advantages to commercial development that focuses more on marketable products, ecology integration, and practical problem solving. In addition, they need to find appropriate application scenarios to industrialize and commercialize AI technologies from labs. For example, in health care, the layout and application of intelligent treatment technology has already taken shape. IBM Watson is applied to clinical diagnosis and treatment and has been promoted in many hospitals in China since 2016; AliHealth is focusing primarily on building intelligent radiology platform; Tencent launched Tencent Miying in August 2017, and which can assist doctors in screening esophageal cancer.

However, the large amount of shared data required by AI poses an "island dilemma" for the data of hospitals and patients. While it is mandatory to break the barriers of all parties, the challenge of how to ensure data security also arises. This situation hinders the real surge of AI in the medical area.

• Confronting the Transformation Difficulties

To objectively understand the development status of the AI industry is to give better play to the enabling role of AI technology. With the prevalence of the digital economy, AI technology has become the innovation power and source of more enterprises, and a basic consensus has been reached about the application of AI in enterprises. But where and how specific applications should be implemented are the key questions to ask in the next stage of AI development.

AI can help enterprises to improve efficiency, but high-efficiency alone cannot help enterprises to form unique competitiveness. In other words, the difficulty in the development of AI market lies in the matching of internal resources and external environment.

We can say that the application of AI technology is the inevitable result of the development of the digital economy business model. Looking back on the history of AI, we will see that AI algorithms such as search recommendation, facial recognition, and speech recognition have achieved rapid growth thanks to the rise of digital economy business models.

If a company relies on data and algorithms to provide services, it means that it must also apply AI technology to develop its own unique competitiveness and bring a better user experience and higher commercial success.

In addition, technology is the premise of further developments in the AI industry, and the premise of technology R&D, landing, and promotion is the full cooperation of top talent in various fields. In the process of promoting the AI industry, the leading role of AI top talent is particularly important.

Unfortunately, China is extremely short of talent in the field of AI. Employees of China's AI industry are mainly concentrated in the application layer, while the talent reserve at the basic layer and technology layer—especially on the processor/chip and AI technology platform—is relatively weak. What's more, the supply and demand of AI talent is seriously unbalanced. Over the past three years, the number of job seekers in the AI field in China has been doubling each year. For AI positions at the basic level (i.e., algorithm engineers), the supply growth has increased by more than 150%. Yet despite such positive expansion, it is still difficult to meet market demand since the time and cost of training qualified AI talent is much higher than that of general IT talent. Thus it is difficult to effectively fill the talent gap in the short term.

To be specific, capital and service are the two main sources for the plight of AI market. In recent years, capital has helped the expansion of AI market and amplified the AI scene effect. It has also enabled technology to realize and seize wealth, which has exacerbated the division among various fields in the AI market.

Today, with legislation and public attention on privacy and data security, the production relationship of productivity services has gradually rationalized, and AI development has also returned to focusing on its essential quality of being an advanced form of productivity.

In this process, Internet enterprises play an important role. As innovators and practitioners of the digital economy, Internet enterprises have created and accumulated a large amount of data in production and business activities. This data came from the real needs, feedback, and behavior of users. On the basis of security and compliance, Internet companies not only make full use of the value of the data but also activate data awareness for the whole business society. Thus, to a certain extent, they facilitated the completion of the big data accumulation for the Third AI Revolution.

But then, with the digital transformation of the whole society, comes the question of how to extend the enabling effect of AI to all aspects of the society. When AI returns to its essence as a technology resource, we should not only have reasonable expectations for it from the perspective of the market and bridge the imbalance between talent supply and demand, but also truly create an AI network that connects data accumulation, technology spillover, and algorithm innovation with different industries. Only thus can we meet the requirement of more scenarios with high frequency, rigid demand, and strong replicability that provide more revenue recognition mechanisms and ensure the landing of the third upsurge of AI.

6.3 Entry and Layout of Global Business Giants

On January 10, 2020, the Key Laboratory of Big Data Mining and Knowledge Management of the Chinese Academy of Sciences released the 2019 *White Paper on the Development of AI.* The white paper analyzed the key technologies and industrial applications in various subfields of AI, and it underlined eight major technologies including computer vision, natural language processing, cross-media analysis and reasoning, and intelligent adaptive learning. It acknowledged the typical AI application scenarios in security, finance, retail, transportation, and education, affirmed the contribution of open, innovative AI platforms in promoting these industries, and posted the list of top-20 global AI enterprises.

• Microsoft: AI Begins with Dialogue

AI has been a strategic goal of Microsoft ever since it was established in 1198.

Microsoft proposed the idea of "dialogue is platform" at the 2016 developer conference. Microsoft believes that the human-computer interaction based on dialogue will replace the keyboard, mouse, and display and become the interface between people and the information world in the future.

Dialogue AI has two main demands: to complete tasks at higher efficiency, and to communicate emotions. These are what Microsoft's conversational AI products are oriented

Top 20 global AI enterprises

Ranking	*Name for the enterprise*	*AI technology*	*Fields applied*	*Country*	*Dates of establishment*	*State of the capital market*	*Market capitalization/ Appraisement/Financing volume*
1	Microsoft	CV, natural language processing, etc.	office	US	1975	publicly listed	market value US$1.21 trillion
2	Google	CV, natural language processing, etc.	comprehensive	US	1998	publicly listed	market value US$932.4 billion
3	Facebook	face recognition, DL, etc.	social	US	2004	publicly listed	market value US$593.4 billion
4	Baidu	CV, natural language processing, knowledge graph, etc.	comprehensive	China	2001	publicly listed	market value US$43.8 billion
5	Da-Jiang Innovations	image recognition, intelligent engine, etc.	drone	China	2006	strategic financing	valued at US$21 billion
6	SenseTime	CV, DL	security	China	2014	Series D round	valued at US$7 billion
7	Megvii	CV, etc.	security	China	2011	Series D round	valued at US$4 billion
8	iFlytek	intelligent speech, etc.	comprehensive	China	1999	publicly listed	market value US$10.8 billion
9	Automation Anywhere	natural language processing, unstructured data recognition	enterprise management	US	2003	Series B round	valued at US$6.8 billion
10	IBM Watson	DL, adaptive learning technology	computer	US	1911	publicly listed	market value US$119.8 billion
11	Songshu AI	adaptive learning technology, machine learning	education	China	2015	Series A round	valued at US$1.1 billion

(Continued)

Ranking	*Name for the enterprise*	*AI technology*	*Fields applied*	*Country*	*Dates of establishment*	*State of the capital market*	*Market capitalization/ Appraisement/Financing volume*
12	ByteDance	Cross-media analysis and reasoning technology, DL, natural language processing, image recognition	information	China	2012	Pre-IPO round	valued at US$75 billion
13	Netflix	video and image optimization, video and image individualization	media and content	US	1997	publicly listed	market value US$141.8 billion
14	Graphcore	smart chip, machine learning	chip	UK	2016	Series D round	valued at US$1.7 billion
15	NVIDIA	smart chip	chip	US	1993	publicly listed	market value US$145 billion
16	Brainco	Brain-computer interface	education, health care, smart hardware	US	2015	Angel round	financing volume US$6 million
17	Waymo	auto drive	transportation	US	2016	Series C round	valued at US$105 billion
18	ABB Robotics	robotic and automation	robots	Switzerland	1988	publicly listed	market value US$51.4 billion
19	Fanuc	robotic	manufacture	Japan	1956	publicly listed	market value US$36.2 billion
20	Preferred Networks	DL, machine learning	IoT	Japan	2016	Series C round	valued at US$2 billion

toward. The smart assistant Cortana on the task-completion end has integrated Windows 10 and become an efficient, inter-platform personal intelligent assistant. The voice assistant Xiaoice has also evolved to a fourth generation since its launch in May 2014. By 2016, the robot "opinion leader" on Sina Weibo and the robot "anchor" of Dragon Television has more than 42 million Chinese users alone.

Microsoft has a unique cognitive computing service. On Microsoft Smart Cloud, cognitive computing has become a common technology module where developers can simply and quickly add intelligence to applications using program interface calls. At present, this cognitive service includes 35 APIs in five categories: vision, language, voice, search, and knowledge, and it is still updating continuously. In addition to software and services that can be invoked on the Smart Cloud, Microsoft also plans to virtualize hardware on the basis of the Cloud so that users can directly obtain computing powers and the hardware pressure can be reduced to a certain extent.

In September 2016, Microsoft merged the Technology and R&D Department and the AI Research Department to form the Microsoft AI and Research Division. As one of Microsoft's strategic core departments, it not only promotes the in-depth integration of AI with Microsoft's own products (i.e., Microsoft Bing, Windows, Office, Xiaoice, Cortana) but also shoulders the responsibility of promoting AI popularization through Smart Cloud and building a common AI platform and system. Microsoft Dynamics 365, for example, a new generation Cloud intelligent business application, helps enterprises grow and digital transformation through perfect integration of CRM & ERP.

The changes in the retail industry are unprecedented. The transformation from traditional physical stores to Cloud stores and Cloud supply chains has essentially changed the ideology of the retail industry. Microsoft's intelligent retail solution helps enterprises improve in Cloud platform construction, Cloud supply chains, and customer service systems and realize the new retail ecology of online-offline integration.

In education, Microsoft Research Asia has developed Xiaoying, a free English learning tool that combines English learning with AI and contains teaching in speaking, listening, vocabulary, Chinese-English translation, and other skills. It has recently launched two new products: Aim Writing and Xiaoluo. The first is a tool platform that helps Chinese learners improve their English writing skills. The second is an English learning app that aims to help Chinese children learn English at home.

• Google: the Absolute AI Giant

Google has more computing power, data, and talent than any company in the world, which naturally makes it a global giant in pursuing AI technology.

Google has more than one billion users for all products it operates: Android, Chrome, Drive, Gmail, Google App store, Google Maps, Google Photos, Google Search, and YouTube. Basically, users relying on Google products can be found anywhere with an Internet connection.

Google's AI development is inseparable from several of its research-oriented AI departments, including Google Brain and DeepMind, acquired in 2014. From a technical point of view, in the field of machine learning algorithms, Google's unsupervised learning had some significant progress in 2020. For example, Google has developed a self-supervised and semi-supervised learning technology called simCLR (a Simple Framework for Contrastive Learning of Visual Representations) that can simultaneously maximize the consistency between different transformation views of the same image and minimize the consistency between transformational views of different images.

On AutoML, Google adopted a search space composed of original operations (addition and subtraction, variable assignment, and matrix multiplication) from the learning code operation of AutoML-Zero in order to deduce modern machine learning algorithms from scratch. In the field of machine perception—how machines perceive and understand the multimodal information of the world, Google presented algorithm models such as CvxNet, deep implicit function of 3D shape, neural voxel rendering, and CoreNet, as well as the practical application in outdoor scene segmentation, 3D human body shape modeling, and image and video compression.

Thanks to the improvement of machine learning algorithms, complex NLP technology can be run on mobile devices to realize more natural dialogue functions. For example, based on the neural network model Transformer, Google created a dialogue robot Meena in 2020 that can process natural dialogues almost at any level. In addition, there are also things like using duplex technology to call companies to confirm whether they are temporarily closed during the epidemic, to achieve three million updates of business information worldwide, and more than 20 billion information displays on maps and searches.

Through the upgrading of machine translation and speech recognition technology, Google Assistant can read web pages out loud in 42 languages and make it easier for users to access web pages. With the help of multi-lingual transmission, multi-task learning, and other technologies, Google has improved the translation quality of more than 100 languages by five BLEU points, which can make better use of monolingual data to improve low-resource languages and provide translation for people from ethnic minorities.

The influence of Google AI goes far beyond the company's product range. External developers now use Google AI tools for all kinds of tasks, from training smart satellites to monitoring changes on the earth's surface to eradicating language attacks on Twitter. There are millions of devices using Google AI at the moment, and this is only the beginning. Google is about to achieve the so-called "quantum hegemony." This new type of computer will be able to perform complex operations at a speed one million times faster than ordinary computers, and will further bring mankind into the age of "rocket computing."

In 2020, Google verified the new quantum algorithm and performed accurate calibration on the Sycamore processor to show the advantages of quantum machine learning or test quantum enhanced optimization; Through the QSIM simulation tool, a quantum algorithm with up to 40 qubits was developed and tested on Google Cloud. Next, Google will establish a general error correction quantum computer according to the technical roadmap to prove that quantum error correction can play a practical role in reality.

• Baidu: to AI Industrialization

Baidu's AI layout starts from building a platform, creating an open ecology, and forming a positive cycle of computing power, scene applications, and algorithms.

The most widely used Baidu AI is the Apollo platform, which covers the fields of intelligent information control, intelligent public transportation, automatic driving, intelligent parking, intelligent freight, and intelligent car couplets. As the world's first open source platform and ecosystem for autonomous driving, Baidu Apollo has successively launched seven versions, bringing 177 ecological partners together through open sources. Now, more than 36,000 developers in 97 countries are using Apollo open source code.

By March 2020, Baidu Apollo has successively cooperated with Changsha, Baoding, Cangzhou, Xiong'an, Chongqing, Hefei, Yangquan and other cities of China in intelligent transportation. Many orders related to the coordinated planning and construction projects have been signed in order to help the local and national construction of intelligent transportation and intelligent cities in China.

According to the "2019 White Paper on AI Development" issued by the Chinese Academy of Sciences, Baidu as a national new generation open, innovative AI platform has built a complete industrial ecology and has become the only enterprise in China with the abilities of automatic driving and vehicle-road collaborative R&D. Baidu Apollo's applications such as limited scene automatic driving, open scene automatic driving, vehicle-road collaborative intelligent transportation, relying on natural language processing, computer vision, and machine learning will effectively improve some of the existing concerns with transportation and traveling.

In the field of input method, the Baidu input method has ranked first in the industry in terms of market share and active user number. The voice input capability of the Baidu input method continues to push limits as it becomes the number one input method product with an average voice request of more than one billion times per day. It has achieved 98.6% speech recognition accuracy, offline Chinese-English "free talk," and dialect "free talk," and similar breakthroughs. At present, it is the third-party mobile phone input method with the highest voice input penetration rate; AI functions such as voice input and handwriting input have a high degree of user recognition. The accuracy of handwriting recognition is now up to 96%, ranking first in the industry, and the accuracy of AI input has exceeded the highest level of the industry by 15%.

With Baidu Xiaodu Assistant, the app store has provided up to 4,300 functions, and the number of developers has reached 45,000 by September 2020. The usage scenarios have also expanded from families, hotels, and automobiles to mobile scenarios. According to the data of Canalys, in the first half of the year 2020, the shipment volume of Xiaodu intelligent speakers ranked first in China; in the first three quarters, Xiaodu intelligent screens were the best seller; during the "618" (June 18) and "double 11" (November 11) shopping festivals, Xiaodu won double sales champion for smart speakers and smart screen category in the whole platform.

In the field of the smart city, with its independent innovation infrastructure complex, including perception center, AI center, data center, knowledge center, and urban intelligence interaction center, Baidu Smart City Solution helps a city improve its level of intelligence and supports public safety, emergency management, intelligent transportation, urban management, and smart education. At present, this solution has been applied in 10 + provinces and cities including Haidian (Beijing), Chongqing, Chengdu, Suzhou, Ningbo, and Lijiang.

In the field of digital finance, the Baidu Intelligent Cloud has served nearly 200 financial clients that include six state-owned banks, nine joint-stock banks, and 21 insurance institutions, and has covered more than ten financial divisions such as marketing and risk control. It has built an ecosystem of more than 30 partners and ranks among the best in the field of financial Cloud solutions in China.

In the field of the industrial Internet, Baidu industrial Internet helps enterprises and up-stream and downstream industries realize digitization, networking, and intelligence. It provides intelligent manufacturing solutions that cover 14 industries, more than 100 enterprises, 30 partners, and 50 vertical scenes and land in 3C, automotive, steel, energy and other fields.

In the field of an intelligent office, Baidu announced Infoflow in May 2020, which relies on the "AI middle platform" and "knowledge middle platform" to build the office assembly line and a new generation AI office platform. Baidu Infoflow enables enterprises to realize intelligent transformation with AI, realizes all-round and intelligent support for the enterprise working mode, and provides full scene services from individuals to organizations, from business to operation.

- **SenseTime: Building an "Urban Visual Center"**

SenseTime ranks sixth in computer vision technology and DL technology, with a valuation of US$7 billion.

Shangtang's AI technology has wide application coverage. It involves not only the traditional fields of security but also the whole smart city sector, such as urban management, smart government, transportation, airport, campus, community, etc. SenseTime's intelligent AI vision platform is a government-authorized new generation of AI open innovation platform.

SenseTime advocates building an "urban visual center" for the city. By opening up the whole link from data acquisition and annotation, model training and deployment, and the upper line of the business system, a closed loop of diversified scenario requirements and efficient model production can be constructed. At the same time, it enables the customer's local model production capacity to independently meet the needs of long-tail development. Finally, it makes contextualized, large-scale, and automatic production of AI algorithms possible.

Based on the "urban visual center," SenseTime launched the "AI City end edge Cloud integration scheme." The project integrates the innovation of the end edge Cloud intelligent full technology stack, provides the central capability of a SenseTime Smart City AI ecosystem, supports the business innovations of the whole scenes of a smart city, and is applied to urban streets, parks, campuses, communities, office buildings, banks, airports, and subways, which affects all aspects of people's life.

In addition, SenseTime also jointly established an urban intelligent ecology with its partners. By connecting upstream and downstream partners, it infiltrates the urban scenes and provides more flexible, more adaptive solutions for urban managers and participants. Various applications and services will be provided, and the three intelligent application systems of urban public services, urban industrial services, and urban people-friendly services will be connected into a comprehensive network that covers more urban scenarios.

So far, the smart city project participated by SenseTime has covered more than 30 provinces, cities, and autonomous regions in China.

In terms of video access volume of a single system, there are several of SenseTime products ranking among the best in China. This requires a high accuracy for algorithms, algorithm diversity, concurrency, and availability of the system. With the "urban visual center" landing in the first-tier cities such as Beijing, Shanghai, Guangzhou, and Shenzhen, the access volume of a single system of more than 100,000 Channels has been realized.

At present, SenseTime is equipped with a professional service team and has established six service centers in China to gradually strengthen its service capacity. At the same time, innovation experiment, evaluation and certification teams are also formed to continuously iterate over data, products, and technologies, continuously optimize and extend core competencies and intelligent applications, and to truly promote the sustainable development of cities.

- **iFlytek: AI security layout**

iFlytek has had a research foundation in AI technologies for many years. It has realized a number of source technological innovations at the core technology level and has made valuable achievements in various international evaluations such as machine translation, natural language understanding, image recognition, graphic recognition, knowledge atlas, machine reasoning, and so on. iFlytek has actively combined its AI core technology with the security industry and shows great momentum in the market.

First, from the perspective of the application of intelligent voice technology, iFlytek's products are applied in voice command and dispatching, voice information releases, and police situation voice analysis, which creates a horizontal command and dispatching mode. On the one hand, it has realized the police situation process of voice access equipment, police resources, voice police dispatches and voice feedback; On the other hand, it achieves one key command and unified dispatching, simultaneous issues alarm information and dispatching instructions, real-time tracking and a timely closed-loop. It greatly improves the efficiency of command, dispatches, police information disposal, service management, and law enforcement supervision.

Second, iFlytek has a core competence platform that includes a big data platform, a Cloud computing platform, and an AI platform. It supports the intelligent analysis center of Xueliang projects and intelligent transportation projects, provides more economical resource utilization, faster computing speed, and higher analytical ability. iFlytek also has image recognition and video structuring algorithms that can analyze the front-end data, correlate the eigenvalues, realize the peer-to-peer application of event detection and vehicle comparison, and provide effective data basis for control deployment.

Third, iFlytek has mature software and hardware systems such as Traffic Super Brain and Whistle Capture. The first is an intelligent application platform that improves the comprehensive practical level of traffic police in traffic management, urban governance, and public service. The application of it in the city of Hefei, Tongling, and Taihe County so far has achieved remarkable results. The Whistle Capture uses core technologies such as interference elimination technology based on deep neural networks and high-precision synchronous multi-source positioning technology, which can adapt to noisy working conditions and maintain accurate positioning. It is mainly used in schools, governments, and other places where whistleblowing is prohibited.

Finally, iFlytek has a multi-dimensional security technology console that can provide "AI middle platform" services for front-end integrators. Examples include whether to wear safety helmets, illegal intrusion identification, fire identification, smoke identification, emergency duty, and an emergency knowledge base. In cases of emergency disposals, for example, when a commander uses the business duty receiving system to communicate with a superior and a subordinate, the system can automatically convert the voice of both sides of the call into text in real time, and realize the intelligent filling of case information messages such as special reports, express reports, and continuation reports through natural language understanding, key element capture, and other technologies.

CHAPTER 7

The Great Powers' Game: Competition, Cooperation, and Governance

7.1 Global Blueprint of AI

As a leading force in the Fourth Industrial Revolution, the AI industry has entered the accelerated development stage. Countries and international organizations including the US, China, the UK, Germany, Japan, France, and the EU are now arranging their AI blueprints from a strategic perspective, by strengthening top-level design, establishing special institutions to comprehensively promote AI implementation, issuing major scientific and technological R&D projects, encouraging funds and financial support, and guiding private enterprises to invest capital resources in the field of AI.

North America, East Asia, and Western Europe are now the most active regions for AI development. The developed countries are privileged in the following aspects: AI theory, technology, talent, and industrial foundations. The US, the EU, the UK, and Japan have had a huge and long-lasting investment in cutting-edge fields such as robotics and neuroscience, and they have successively released national robot plans, human brain plans, and autonomous system R&D plans (e.g., autonomous driving). In 2016, the US first released the "National Strategic Plan for AI R&D," and many countries followed her step in including AI management into their national strategic plan: 2017—Japan, Canada, and the UAE; 2018—the EU, France, the UK, Germany, South Korea, and Vietnam; 2019—Denmark and Spain.

In summary, to guide AI innovative development with strategic means is becoming a new global trend. Spontaneous and decentralized research is gradually developing to fit into the mode promoted and led by national strategy and themed by industrialization and application.

• United States: Maintaining Global Leadership in the AI industry

The United States is a superpower that leads the global AI management. In recent years, a series of policies, bills, and promotional measures have been issued to support the advanced blueprint in neuroscience, quantum computing, and general AI based on rich research achievements. At the same time, with the tremendous advantage of Silicon Valley, the complete AI industrial chain and ecosystem led by enterprises enables US global leadership in basic software and hardware industries, such as AI chips, open source framework platforms, and operating systems.

During the Obama Administration, the US government actively promoted the development of AI and supported its basic and long-term development. In the second half of 2016, the government released three reports with global influence: "Preparing for the Future of AI," "National Strategic Plan for AI R&D," and the "Report on AI, Automation, and the Economy." These documents provided advice respectively on the development of AI for US Federal Government and relevant institutions, US AI R&D, and AI's impact on the economy.

In February 2019, President Donald Trump signed an executive order to launch the American AI Initiative that mobilized more federal funds and resources at the national level to invest in AI research focusing on five areas: R&D, resource openings, policy-making, talent training, and international cooperation. In 2019, the US updated the "National Strategic Plan for AI R&D" and identified eight strategic directions for priority development: basic research, man-machine cooperation, ethics, law and social impact, security, public data and the environment, standards, human resources, and public-private cooperation. It strengthened policy formulation and investment guidance, increased long-term investment in defense science and technology, and emphasized the importance of public-private cooperation.

"Understanding and addressing the ethical, legal, and social impact of AI" is included in the ethics strategy in the plan. It requires studying the expression and coding of human values and belief systems, creating an ethical AI, formulating an acceptable moral reference framework, and realizing the comprehensive design of an AI system that conforms to the ethical, legal, and social objectives.

In 2018, the US established the National Security Committee on AI, which was responsible for investigating AI ethical issues in the domains of national security and national defense. Meanwhile, the US has introduced AI ethical education into the talent training system. Since 2018, Harvard, Cornell, MIT, Stanford, and many other universities have opened interdisciplinary courses on AI ethics.

In February 2019, President Trump issued the executive order on "Maintaining the Leadership of American AI" with a special focus on ethical issues. It required the US to cultivate public trust and confidence in AI technology, protect civil liberties, privacy, and American values in application, and fully tap the potential of AI technology.

In June 2019, the National Science and Technology Council of US issued the "National Strategic Plan for AI R&D" to implement the above-mentioned administrative orders. It proposed that AI systems must be trustworthy, and an ethical AI system should be designed with higher justice, transparency, and accountability.

In August 2019, the National Institute of Standards and Technology (NIST) issued a guidance on AI technology and ethical standards, which required that the standards should be flexible, strict, and timely. It confirmed the improvement of justice, transparency, and responsibility mechanisms to design an ethical AI structure in line with moral, legal, and social goals.

In October 2019, the Defense Innovation Commission of the US launched the "Principles of AI: Some Suggestions on the Application Ethics of AI from the Department of Defense" and put forward the five principles of "responsibility, justice, traceability, reliability, and controllability" for the design, development, and application of AI technology in combat and non-combat scenarios.

The US government also places importance to the impact of AI on employment. The House of Representatives issued the AI Innovation Team Act in 2017 and the AI Employment Act in 2018, proposing that the US should create a lifelong learning and skill training environment to meet the challenges brought by AI to employment.

The Autopilot Act was passed in the House of Representatives in 2017, and the "Preparing for Future Traffic: Autopilot 3" was issued by US Department of Communications in 2018. The Department of Health and Human Services also issued a *Data Sharing Manifesto* to regulate and manage automatic driving vehicle design, production, and testing so as to ensure user privacy and safety.

• China: A National Strategy and New Infrastructure Building

AI is an important technology for China to deepen the supply-end-reform and promote the development of a digital economy. AI development in China is the manifestation of the Central Committee of the Communist Party of China (CCCPC) and the State Council taking advantage of the general trend of the current scientific revolution and industrial reform. Relevant departments in China are placing great emphasis on guiding the healthy development of AI in order to seize the major strategic opportunity to accelerate the construction of an innovative country and a world superpower in science and technology.

In China, governmental support for AI development varies in form. A joint promotion mechanism involving the Ministry of Science and Technology, the National Development and Reform Commission, the Central Network Information Office, the Ministry of Industry and Information Technology, and the Chinese Academy of Engineering was formed. Since 2015, the Chinese government has issued multiple policies providing a large number of funds for the development and implementation of AI technology as well as tremendous support for the introduction of AI talent and enterprise innovation. These policies not only offered political guidance but also positive signals to the capital market and industry stakeholders.

The years of 2015–16 were the initial period for AI policy formulation. It mainly focused on system design, technology R&D, and standard formulation that laid the technical foundation for follow-up development.

In July 2015, the State Council issued the "Guiding opinions on actively promoting the 'Internet plus' action," which incorporated AI into one of the key national tasks. It marked the formal opening of the period of special AI in formulating industrial policies.

In May 2016, the NDRC issued the "'Internet plus': A Three-year Action Plan for AI" that proposed to build AI basic resources and innovation platforms, establish an AI industrial system, innovate service systems and standardization systems, break through the core technology, and cultivate some leading AI enterprises.

In August 2016, the State Council issued the "Thirteenth Five Year Plan for National Science and Technology Innovation," which clearly regarded AI as a major science and technology project reflecting the national strategic plan.

In March 2017, "Artificial Intelligence" was written into the national government work report for the first time.

In July of the same year, the State Council issued the "Development Plan for a New Generation AI," which officially marked AI's national strategic position at a comprehensive level. It was China's first system deployment document in the field that specifically planned and deployed the overall idea, strategic objectives, tasks, and safeguard measures for the development of a new AI in China until 2030. It listed different expectations for AI's basic layer, technology layer, and application layer, and it established three-step goals for 2020, 2025, and 2030:

By 2020, China's AI technology and application will keep pace with the world's advanced level standard, and the industry will become a new important economic growth point with the scale of its core industry exceeding CN¥ 150 billion and driving the scale of its related industries to over CN¥ 1 trillion.

By 2025, major breakthroughs will be made in the basic theories of AI, and some technologies and applications will reach the world's leading standards. AI will become the main driving force for China's industrial upgrading and economic transformation. The scale of core industries will exceed CN¥ 400 billion, and the scale of related industries will exceed CN¥ 5 trillion.

By 2030, AI theory, technology, and applications will generally reach the world leading standards and become the world's main AI innovation center. The scale of core industries will exceed CN¥ 1 trillion and drive the scale of related industries to more than CN¥ 10 trillion.

In October 2017, AI was included in the report of the Nineteenth Report of the National People's Congress to "promote the in-depth integration of the Internet, big data, AI, and the real economy."

In December 2017, the Ministry of Industry and Information Technology issued the "Three-year Action Plan for Promoting the Development of the New Generation AI Industry (2018–2020)" that specified the indicators of multiple tasks for the next three years from the perspective of promoting industrial development, refined and implemented the relevant tasks, and promoted AI industrialization and integrated applications through deep integration of information technology and manufacturing technology. At the same time, upgrading and intellectualization of traditional industries were strongly encouraged. Important documents such as *The Intelligent Manufacturing Development Plan (2016–2020)* and *The Industrial Structure Adjustment Guidance Catalogue (2019)* were also issued to provide a strong policy guarantee for industrial upgrading.

In January 2018, the *White Paper on AI Standardization* (2018 Edition) was officially released, and the standardization work entered the stage of comprehensive overall planning and coordinated management. In March, AI was once again included in the government work report that emphasized the industrial AI applications. In areas of health care, pension, education, culture, sports, and others, "Internet plus" was promoted to develop an intelligent industry and expand intelligent life.

In May 2018, Chairman Xi Jinping stated at the Two Academies' Conference: "The new round of technological revolution and industrial transformation are reconstructing the global innovation map and reshaping the global economic structure. Science and technology have never had a profound impact on the future and destiny of the country as it is today… Now, we have ushered in a new round of scientific and technological revolutions and a historical period for China's development mode. We are facing a once-in-a-lifetime opportunity, but we are also facing the severe challenge of a widening gap."

In October 2018, Chairman Xi organized the Ninth Collective Learning of the Political Bureau of the CCCPC on the topic of AI. The conference focused on building a modern economic system, giving play to the role of AI in quality changes, efficiency changes, and dynamic changes, strengthening the combination of AI with the protection and improvement of people's livelihood, and creating more intelligent working and lifestyle tasks.

In November, the Ministry of Industry and Information Technology issued the "Work Plan for Unveiling the Key Tasks of Innovation in the New Generation AI Industry" focusing on the key directions of "cultivating intelligent products, breaking through the core foundation, deepening the development of intelligent manufacturing, and building a support system." A number of departments with key core technologies and strong innovation abilities were selected to tackle key technical problems.

2019 was the third year that AI appeared in a government work report. The report proposed to promote transformation of traditional industries and to build an industrial Internet platform for upgrading AI to "Intelligence plus." Emerging industries need strong support, and the digital economy needs to be empowered by thorough R&D and the application of big data and AI technologies.

In March 2019, at the seventh meeting of the Central Comprehensive Deepening Reforms Commission, the "Guiding opinions on promoting the in-depth integration of AI and the real economy" were adopted. The opinions emphasized grasping the developmental characteristics of the new generation AI technologies, exploring the transformation path of the innovation achievements in combination with the regional characteristics of different industries, and building a data-driven smart economy featuring man-machine collaboration, cross-border integration, and co-creating and sharing.

In the "Guidelines for the Construction of National New Generation AI Innovation and Development Pilot Zones" issued by the Ministry of Science and Technology in August 2019, it was proposed that by 2023, around 20 pilot zones will be constructed and multiple goals will be accomplished. These include: innovating practical and effective policy tools, forming typical modes of deep integration of AI and economic and social development, accumulating replicable and generalizable experiences and practices, and creating AI innovation high ground. Construction of a number of national AI innovation pilot zones were approved in Beijing, Shanghai, Tianjin, Shenzhen, Hangzhou, Hefei, Deqing County, Jinan, Xi'an, Chengdu, Chongqing, and other places.

The COVID-19 pandemic started in the beginning of the year 2020, and accelerated the pace of new infrastructure development. Today, AI has become one of the seven pioneers in the industry as well as the general basic technology in the information field that provides a series of chips, devices, algorithms, software frameworks, and platforms. Apparently, promoting the AI new infrastructure will be beneficial to the intelligent upgrading of traditional industries, which will in turn promote the upgrading and evolution of AI technology.

In the long run, as an integrated innovation tool to support the transformation and upgrading of traditional infrastructure, AI will play a major role in completing the digital transformation and intelligent upgrading of the industrial chain, realizing the efficient allocation of industrial factors, and helping the growth of new kinetic energy for economic development. With the continuous political support, it is foreseeable that the prospect of China's AI industry is a positive one.

• European Union: Ensuring the Global Competitiveness of AI in Europe

In order to promote the common development of AI in Europe, the EU has been actively promoting an AI cooperation program. Between April and July 2018, 28 EU member states jointly signed the "Declaration on AI cooperation," promising to form joint forces in the field of AI and carry out a strategic dialogue with the European Commission. By the end of the year 2018, the EU issued the "Collaborative Plan on the Development and Use of AI in Europe," which proposed to take joint actions to promote cooperation among EU member states, Norway, and Switzerland in the following four key areas: increasing investment, providing more data, cultivating talent, and ensuring trust. In January 2019, the AIFOREU project was launched to establish demand and open a cooperation platform. It integrated and gathered AI resources (i.e., data, computing, algorithms, tools) from 79 R&D institutions, small- and medium-sized enterprises, and large enterprises in 21 member states and provided unified and open services.

In order to ensure the global competitiveness of European AI, the EU took many measures. These included issuing the "EU AI Strategy," signing the cooperation declaration and collaborative plan, jointly arranging R&D and applications, ensuring a people-oriented development path, building a world-class AI research center, and carrying out pioneer research in neuroscience, intelligent society, ethic studies, and other fields.

The EU places a specific emphasis on establishing an ethical and legal framework of AI that ensures the development of AI technology is beneficial to individuals and society. In the "EU AI Strategy" report released in April 2018, the establishment of an appropriate ethical and legal framework was made one of the three strategic priorities. The AI High Level Group (AIHLG) was formed to draft an AI ethics guide.

In April 2019, the AIHLG published the "Ethical Guide to a Trusted AI," which introduced the concept of "trusted AI" for the first time. Based on the core value of "united in diversity," the organization pointed out that trust is the cornerstone of a society, community, economy, and sustainable development within the context of rapidly changing science and technology. The EU believes that only with a clear, comprehensive, and trustworthy framework can humans and communities have confidence in technological development and its applications, and only through credible AI that is consistent with their values (i.e., respect for human rights, democracy, law and order) can European citizens obtain benefits from the technology.

Trusted AI is composed of three elements: legitimacy, ethics, and robustness. It is also constructed upon a three-tier framework: four basic ethical principles—seven basic requirements—Trusted AI evaluation list. The framework raised specific and practical evaluation criteria both for enterprises and regulators.

Once again, in the *White Papers on AI: Europe's Path to Excellence and Trust*, the EU proposed that winning people's trust in digital technology is the key to technological development. The

EU has created a unique "trust ecosystem" based on European values and basic rights such as human dignity and privacy protection.

Due to the cultural and language differences, European countries are subjected to the challenge of forming big data sets. Nevertheless, with an advantage in the construction of the global AI ethical system, the EU occupies a leading position in AI technology regulation.

- **United Kingdom: Building a World-class AI Innovation Center**

In recent years, the British government has released a series of policies that centered on an active promotion of industrial innovation and the country's leading position in AI ethics, supervision, and governance. The aim of such governmental support on AI industry is to make the UK the world innovation center and once again lead the development of the global science and technology industry.

For example, in "The Industrial Strategy: Building a UK Adapted to the Future" released in 2017, the British government established four priority areas for AI development: build the UK into a global AI and data Innovation Center; support various industries to use AI and data analysis technology; maintain a leading position in data and AI security; cultivate their citizen's work skills.

In April 2018, in "Action in the Field of AI" jointly published by the Department of Business, Energy, and Industrial Strategy (BEIS) and the Department of Digital, Culture, Media, and Sports (DCMS), it was proposed to invest in R&D, skills, and regulatory innovation; support industries to improve productivity through AI and data analysis technology; strengthen Britain's network security capabilities.

Several other documents mentioned the raising of governmental funds in R&D and prioritizing support in key areas of studies. The British government will increase R&D investment (including AI technology) to 24% of GDP in the next 10 years. By 2021, the R&D investment will reach £12.5 billion, with £93 million from the Industrial Strategic Challenge Fund allocated for R&D on robot and AI technology. At present, many innovative AI companies have emerged, and the UK government is actively launching incentive policies for start-ups.

In terms of AI ethics, the UK has established a data ethics and innovation center responsible for ensuring that the application of data (including AI) is safe, innovative, and ethical.

The Institute of AI and Machine Learning Ethics in the UK puts forward eight principles of "responsible machine learning" and authorizes all actors (from individuals to nations) to develop AI. These principles involve human control and maintenance, appropriate remedies for the impact of AI, bias evaluation, interpretability, transparency, and repeatability. They also imply reducing AI automation's negative impact on laborers, accuracy, cost, privacy, trust, and security.

• Japan: Building a "Super Intelligent Society" with AI

In January 2016, the Japanese government promulgated the "Fifth Basic Plan for Science and Technology" and raised the "Super Intelligent Society 5.0 Strategy" with AI as the core. In April 2016, Prime Minister Shinzo Abe put forward the objectives of setting an AI industrialization roadmap and establishing an AI technology strategy conference.

Aiming for the construction of "Super Intelligent Society 5.0," Japan identified the year 2017 as the first year of AI and issued a national strategy that comprehensively expounded Japan's AI technology and industrialization blueprint. The strategy expressed the hope to strengthen the country's global leading advantages in the fields of automobiles and robots, as well as to address the national problems in pension, education, and business. There was also the AI Technology Strategy Conference that promoted the R&D and AI applications in industry, academia, and government.

In the year 2017, the Japanese government provided abundant financial support for AI R&D, which many Japanese enterprises have also participated in. Cui Cheng, member of the Macroeconomic Research Institute of the National Development and Reform Commission believes that the AI R&D in Japan has the following important characteristics:

First, place emphasis on top-level design and strategic guidance, and make AI the core of the super intelligent society 5.0 construction. Sun Zhengyi, President of Softbank, also known as "Buffett + Gates" in Japan, put forward the ambitious proposal of saving Japan with AI robots at the end of the year 2014 and to bring Japan's industrial competitiveness back to being the best in the world by 2050;

Second, strengthen the construction of systems and mechanisms. The Ministry of Internal Affairs and Communications (MIC), the Ministry of Education, Culture, Sports, Science and Technology (MEXT), and the Ministry of Economy, Trade and Industry (METI) jointly promote the progress under the mode of tripartite cooperation;

Third, emphasize users with comprehensive coverage. Japan's AI research and development is not only aimed at industrial departments but also at social departments such as transportation, medical health, and nursing regarding people's livelihoods;

Fourth, the operation is government-guided and market-oriented. Research institutions subordinate to the government take the lead in carrying out R&D activities, and jointly promote them by providing subsidies to private enterprises and universities.

Fifth, accentuate local features and advantages. Japan's industrial strengths are in the fields of automobiles, robotics, and medical treatment. Its AI R&D also focuses on these fields. Based on the market demand for AI robots in an aging society, health and nursing, as well as the construction of super intelligent society 5.0, Japan highlights the characteristics of improving software with hardware, and upgrading industries with innovative social demand.

In June 2018, the Japanese government introduced a plan to promote the popularization of AI at the AI Technology Strategy Conference. Right now, Japan is actively publicizing the national AI strategy and industrialization roadmap with the aim to realize the goals mentioned above.

As for basic research and application, the MIC is mainly responsible for studying brain information communication, voice recognition, innovative network construction, etc.; The MEXT is mainly responsible for basic research, new generation basic technology development, and talent training; The METI is mainly responsible for the practicality and social AI applications.

Japan has joined forces with the government, academia, and industry to promote technological innovation and the development of AI industry. As a national comprehensive management organization, Japan's AI Technology Strategy Committee is in charge of facilitating the cooperation among the three Ministries. At the same time, Japanese scientific research institutions also actively strengthen cooperation with the whole industry and vigorously promote the industrialization of AI R&D achievements.

• Germany: Building "AI Made in Germany"

As early as the middle and late 1970s, Germany put forward an AI-relevant policy, the "Plan to Improve Working Conditions." The plan set mandatory provisions for some high-risk jobs, requiring that the dangerous tasks must be carried out by robots. In the following decades, the development of German AI industry has always been closely related to robotics and industrial manufacturing.

At the 2013 Hannover Messe, the German Federal Government publicized *Industry 4.0 Strategy*, which focused on the construction of intelligent chemical plants. The aim was to improve the competitiveness of German industry with new technologies such as the Internet and AI and to take the lead in the new round of industrial revolution dominated by intelligent manufacturing. As the key technology of Industry 4.0, AI is playing an increasingly important role in Germany's national strategy. A series of development plans and research reports related to it have been put forward, including the "High-tech Strategy 2050," the "Annual Evaluation Report on Research, Innovation, and Technological Capability" issued by the German Research and Innovation Expert Committee (EFI), "Discussion on AI strategy" jointly carried out by Germany and France, etc.

Industry 4.0 and advantage in intelligent manufacturing define Germany's AI blueprints, which emphasizes brand building "AI made in Germany" and achieving the highest standard in R&D and AI application.

In September 2018, the German Federal Government issued the "High-tech Strategy 2025." One of its 12 tasks was to "Promote the application of AI and make Germany one of the world's leading centers in researching, developing, and applying AI." The strategy also clearly proposed to establish an AI competitiveness center, formulate an AI strategy, establish a data ethics committee, and establish a German-French AI center. In February 2019, the Federal Ministry for Economic Affairs and Energy (BWWi) issued a draft of "National Industrial Strategy 2030" that repeatedly stressed the importance of this target technology.

In July 2018, the German government released the "Key Points of the Federal Government's AI Strategy" that mentioned building an "AI made in Germany" that can be internationally recognized.

To achieve this goal, in November 2018, the document "Federal Government's AI Strategy" formulated three strategic objectives and twelve areas of action including research, technology transformation, entrepreneurship, talent, standards, institutional framework, and international cooperation. Specific measures proposed in the strategy included: supporting start-ups, building a European AI innovation cluster, developing new technologies closer to small- and medium-sized enterprises, increasing and expanding AI research centers, etc.

7.2 Seizing the AI High Ground

Uncertainty shapes relationships.

The world was caught off guard when the COVID-19 pandemic broke the once linear, smooth, and predictable ways of life. In the era of economic decoupling and geopolitical polarization, science and technology have become the crucial elements affecting international relations in the post-epidemic time. Digital technology represented by AI and big data will fundamentally reshape the fields of health care, education, finance, transportation, entertainment, consumption, national defense, and more.

At the moment, many countries are planning their AI strategies at a national level. The competition on the AI high ground has changed and redefined the rules of global politics and economy, the thinking about the future of mankind, and the process of globalization.

• Why the Need to Seize the AI High Ground?

In September 2017, Russian President Putin declared that "AI (as) the future of mankind … brings enormous opportunities as well as unforeseeable threats. Whoever takes the lead in this field will become the leader of the world."

Apparently, progress in the field of AI determines the future. From an economic point of view, AI is an important driving force for economic growth.

On the one hand, AI gives impetus to the intelligent transformation of industry. On the basis of digitization and networking, it can reshape the mode of production organization, optimize the industrial structure, promote the intelligent transformation in traditional fields, lead the industry to the high end of the value chain, and comprehensively improve the quality and efficiency of economic development.

On the other hand, the popularization of AI will promote multi-industry innovation, which can greatly improve the existing labor productivity and open up a new space for economic growth. According to Accenture's prediction, in 2035, AI will improve China's labor productivity by 27% and create an additional $7.1 trillion to the total economic aggregate.

From the military point of view, in the weapon systems, logistics support systems, and command decision-making systems, where the application of AI is mainly reflected, it can promote the transformation of tactics and soldiers and be widely used in network war. The R&D and deployment of AI in the military field will greatly improve the military strength of the host country and affect the balance of international military strength.

First, the biggest difference between AI weapons and traditional ones is "intelligence" or "autonomy." This is why AI weapons are also called "autonomous weapons" or "autonomous weapon system." It is regarded as the third revolution in the history of human warfare, after gunpowder and nuclear weapons.

Second, AI can generate a battlefield integrated platform based on massive logistics support data, systematically and comprehensively analyze and evaluate various logistics support schemes, and intelligently select the best support scheme. When used to assist command decision-making, it can make key contributions to intelligence reporting, monitoring and investigation (ISR), and analyzing. It can comprehensively restore the battlefield information, simulate force deployments and the combat capability of both sides, and complete relatively accurate battlefield sand table deductions.

Third, AI can also be used for cooperative operations between weapons and between man and machine to enrich tactics. In addition to the traditional battlefield, AI has become a new weapon in modern war. For example, in 2019, IBM introduced a new malware which imitated features of the neural networks to implement targeted attacks on countries and organizations that previously had a large number of computing and intelligence resources.

From the perspective of the acquisition of global political power, the impact of AI is mainly reflected in the differences of countries' ability to obtain and analyze big data resources. Today, the value of data is becoming more and more prominent. Data represents a winning weapon, and it has increasingly become a strategic resource of state power.

Interconnection led to the booming of data. Data collecting and processing will give rise to a number of new technology enterprises that can ultimately affect the development of the world economy and military. Undoubtedly, AI's role in data mining and analysis is indispensable.

Of course, AI is not a weapon like missiles, submarines, or tanks. Instead, it is more similar to internal combustion engines and electricity. It is an enabler and a general technology with a wide range of applications.

• AI Echelon Formations Appear

For all countries, AI is at the heart of their competitive advantage in the new round of international games and the changing international systems.

Different countries have different focuses. In the US, military applications lead the way for technology industry development, which is market-and-demand oriented. Attention is placed on leading global economic growth through high-tech innovation as well as to the formulation of product standards. In Europe, AI development focuses on the R&D environment and the formulation of ethical and legal rules. In Asia, the industrial application demand is the driving force for AI development, and emphasis is placed on R&D in industrial scale and local key technologies.

Right now, China and the US are the two major players in the game of the entry and layout of global AI technology. In comparison, the US has great advantages at the basic layer. Take the open source algorithm platform as an example: while Google, Facebook, and Microsoft have all launched their own open source platforms for DL algorithms, China only has Paddle from Baidu.

With the Cloud platform in the technology layer, the US occupies a dominant market position. But China has also launched leading Cloud service platforms through Internet giants such as Alibaba, Huawei, and Tencent.

In the application layer, both Chinese and US Internet giants have their own vertical application platforms. For example, in terms of voice platforms, Google assistant, Microsoft Cortana, iFlytek voice open platform, and Baidu Brain are all well-known in the industry.

AI is now included in the international agenda. However, new international norms on this world-changing technology has not yet taken shape or produced substantive effects. Although it has challenged our cognition of the established lifestyle and impacted the international competition pattern and situation quite thoroughly, the real power of AI is yet to be seen.

7.3 Technology is Not Neutral

Distinct from any stage of human history, there is a spontaneous and prominent respect for technology in the industrial era, where technology is viewed as the means of rationality. The age of AI has reshaped the relationship between man and technology, which are no longer merely tools to "create" and to be "used" but a humanized natural reality.

In the age of intelligence, the understanding and domestication of technology have become the new focus. With more legal and technical issues arising, the traditional concept of "neutral technology" is receiving more suspicion as well.

• Technology with Purpose

Whether the steam engine appeared in the First Industrial Revolution or the electric power and internal combustion engine in the Second, the essence of technology has always remained unchanged, which is to improve production efficiency and to expand the extension of life. The only difference is that modern technology has a more objective orientation than in the past due to the objectivity of modern science.

For a long time, it has been believed that technology derived from science has no moral bias per se and its neutrality has been confirmed from the three aspects of function, responsibility, and value.

Functional neutrality holds that the mission of technology is to follow its functional mechanism and principle. This concept is particularly reflected through network neutrality. This means that network operators and providers remain neutral to user needs without providing differential treatment in data transmission and information content transmission.

Responsibility neutrality separates the technical function from the practical consequences, which means that technology users and implementers do not need to bear responsibility for the negative effects of technology on society as long as they have no subjective intention. In other words, a kitchen knife can both cut vegetables and kill people, but the producer of kitchen knives should be held responsible for the consequences of murder with kitchen knives.[1]

Ultimately, both functional neutrality and responsibility neutrality lie in the value neutrality of technology. In the Third Industrial Revolution, "neutrality" was not applicable to any technology-related behavior, whether technology assumption, development, communication, application, or regulation. People's values have long been integrated into everything we design and build.

Different from the evolutionary process that follows natural selection, technology is decided by the will of the inventor for a certain purpose. Although technology contains an objective structure, it serves people's purposeful and rational needs. This means that technological

1. A customer may buy a kitchen knife for killing or for cooking, but a salesperson never knows what the kitchen knives he sells is used for.

products are always conceptualized and thoroughly planned when they come out, because all new creations are goal-oriented.

When demand is greater than supply, any product can meet the needs of the Internet consumers. But when the user scale stops growing, technology companies must develop more products in line with commercial value in order to survive. Therefore, technology neutrality will inevitably be affected by business preferences. As Abraham Kaplan describes in the "law of instruments," to someone whose only tool is a hammer, everything looks like a nail that needs to be pounded. "Neutrality" does not fit in the context of commercial value that is based on a profit-driven foundation.

- **Technology Non-neutrality**

The "non-neutrality" of technology is not a recent concept. In 2014, the White House released the strategic white paper "Big data: seize opportunities and preserve value." It emphasized the importance of the "first law of technology," which refers to the idea that "technology has no positive or negative connotation, but technology is not neutral." With the rapid development of big data, the US made risk management of data security the core of the "datacentric" strategy, ensuring "information advantage" and "decision-making advantage" through data "weaponization."

When technology manifests people's technological purposes and determines people's business orientation, technology non-neutrality will be the ultimate result. In fact, AI infringement begins as early as the stage of data collection, which is the starting point for AI technology application.

In the age of intelligence, the real world and the virtual space are connected through Web 2.0, which is also the means for government and enterprises to collect immeasurable data and any trace of network user activities. Unencrypted data makes immeasurable personal information fully unprotected for disclosure and illegal exploitation. This accounts for the first stage in technology non-neutrality.

The boundary between private and public space is getting blurry in ubiquitously concrete aspects through covert micro channels. AI technology has become a knowledge form carrying power in Michel Foucault's value system. Technology innovation is accompanied by the growth of micro power that controls society. It is in the hands of the state but can also be owned by enterprises.

Therefore, in the era of technological innovation and development, the once private information can be collected, copied, disseminated, and used without the knowledge of the information owner. This not only enables the privacy infringement to take place out of any relationships at any time and any place, but also enables enterprises to convert the occupied information resources into business value through data processing and react against users' will and desire through AI media. This accounts for the second stage in technology non-neutrality.

Now, the impact of algorithms on human beings has penetrated into almost all fields of life and can be seen through the decision-making processes of individuals, enterprises, and broader society. Unlike traditional machine learning, DL algorithms do not follow the old paradigm of "data acquisition and input—feature extraction and selection—logical reasoning and prediction." Instead, it can automatically learn according to the most initial features of things and generate higher-level cognitive results.

This means that there is a "hidden layer" between AI data input and output, or the so-called "black box." Described as a linear causal relationship or a calculable exponential increase, the black box becomes "white," meaning that the operation is controllable and the output is predictable.

However, once the black box does not meet people's description, the box remains "black." People must accept the idea that input does not explicitly determine the output, and the system or the black box determines for itself. This is very important. Obviously, future technologies will grow even more powerful and far-reaching than today, and when AI makes its own moral choices, it will be meaningless to adhere to technology neutrality. This accounts for the third stage in technology non-neutrality

Of course, scientific objectivity complements technology reliability, but technology designers or groups also have their own value orientation that influences their commitment to scientific significance. At the same time, designers cannot escape social influence when comprehending scientific significance, which means all technologies are constructed within specific historical and cultural backgrounds and based on specific ideologies and purposes.

The downside of the long-proven non-neutral technology is gradually showing due to the insufficient development of social interpretation systems compared to that of science and technology. Given its intimate interrelationship with human society, technology reminds us of the importance of moral values in the world we are now living in.

• Beneficial AI

There is a classic "tram paradox" in the field of unmanned driving. The reason this paradox is important is that it reflects the diversity of human morality. When a sacrifice of human life must be made in a car accident, we see a variety of choices based on people's moral differences: some choose to protect the passengers, while some choose to protect the passers-by; some choose to protect the elderly first, while some choose to prioritize the children and women.

AI turns the originally scattered, random questions into a fixed problem with an algorithm. More specifically, AI uniformly collects and categorizes moral concepts under the question of "who gets systematically sacrificed?" But the question is, can this result be widely accepted? This creates huge and controversial moral and social problems.

Of course, the "tram paradox" is only one example of AI morality, and the probability of real car accidents is very small. What the "tram paradox" embodies is the profound impact that AI will leave on society. Although the theory and algorithms of AI are becoming increasingly mature, its social influence is still uncertain. Thus the construction of technical values is particularly important in order to give full play to the utility of AI in society.

Generally speaking, the development of AI should focus on a "beneficial AI" that accurately prevents and deals with the possible risks while continuously releasing technology dividends. It should balance AI innovative development and effective governance, to improve the relevant algorithm rules, data use, security, and other governance capabilities in order to create a standardized and orderly developmental environment for AI.

As mentioned earlier, risks that AI technologies can bring include algorithm discrimination, privacy protection, rights protection, and severe challenges such as social unemployment and national security instability. In view of the width and depth of these problems, it is necessary to re-examine ways of accurate prevention through the perspective of global governance so as to avoid the negative impact of AI on human society to the fullest.

For example, the "National AI strategy" of South Korea raises a guideline for healthy AI development. Specific requirements include: "prevent negative effects of AI, formulate AI an ethics system, and promote the construction of a quality management system for monitoring AI reliability and safety."

Balancing the relationship between innovation and development of AI and effective governance is the key. On the one hand, excessively strict governance will limit the progress of AI and results in difficulties in any technological innovation. On the other hand, AI development without any supervision and regulation is easy to "go astray," which will bring harm to human society and deviates from the idea of "beneficial AI."

In sum, it is crucial to adhere to a safe and controllable governance mechanism, pay equal attention to the development of open and innovative technology, and give appropriate trial-and-error space for technological progress and market innovation. We should neither simply and completely stifle AI technology nor allow it to spread without surveillance. We should give full play to the effectiveness of collaborative governance of multiple subjects, enable all parties to perform their duties, and hold the governance bottom line in order to ensure the innovation vitality and development power of the industry as well as to enhance users' sense of gain and security.

7.4 Multi-Subject Participation and Global Collaborative Governance

AI governance started in 2017 in the US, the EU, and the UK (which has initiated Brexit). According to the data from Morning Consult, more than 67% of those questioned explicitly support this idea.

Countries all over the world generally express a positive attitude towards AI governance. For example, the US, Japan, South Korea, the UK, the EU, and UNESCO have successively launched ethical principles, norms, guidelines, and standards in terms of robot principles and ethical standards. In addition, international organizations and industry alliances also play an important role in this process.

• International Organizations and Industrial Guidelines

AI development features transnationalism and an international division of labor. This means that AI governance calls for international organizations to strengthen coordination and cooperation among countries.

On the one hand, intergovernmental organizations guide the formation of consensuses among major countries in the field of AI. Due to the different attention and investment in various countries, the rules on AI governance took the lead in forming and spreading in developed countries. As a weather vane leading the formulation of international rules, intergovernmental organizations will discuss AI regulation based on the principled declarations of different countries, and they will guide AI governance to reach an international consensus.

On the other hand, intergovernmental organizations can promote the global sharing of AI platform management rules. Due to the disequilibrium in the development of AI around the world, many developing and underdeveloped countries have not yet incorporated AI governance into their national strategies. Therefore, the forward-looking release of AI governance rules by intergovernmental organizations will narrow the gap among countries and promote a coordinated, healthy, and shared global environment for AI development.

The United Nations, adhering to international humanitarian principles, put forward the "principle of meaningful human control over lethal autonomous weapon systems" in 2018. It also brought up the initiative that "all lethal autonomous weapon systems that can be separated from human control should be prohibited" and established the Institute of Crime and Justice in The Hague for robotic studies and AI governance.

The Group of Twenty issued the "G20 AI principles" in June 2019, advocating the development of AI with a human centered and responsible attitude. It proposed specific rules including "investing in AI research and development, cultivating a digital ecosystem, creating a favorable policy environment, preparing human capacity for labor market transformation, and realizing reliable international cooperation."

The Organization for Economic Cooperation and Development (OECD) issued the "Intergovernmental Policy Guidelines on AI" in May 2019, advocating inclusive growth and sustainable development of AI that can benefit both people and the earth. According to the document, the design of AI systems should "respect the rule of law, human rights, democratic values, and diversity, and include appropriate safeguards to ensure a fair and just society."

At the G7 Summit in 2018, 12 principles regarding AI were agreed upon. These were: 1) promote people-oriented AI and its commercial application, and continue to promote the adoption of neutral technical ethics; 2) increase investment in AI R&D to publicly test new technologies and support economic growth; 3) provide guarantees for human labor to receive education, training, and learn new skills; 4) provide guarantees for underrepresented populations (i.e., women, marginalized communities) to involve them in the development and implementation of AI; 5) promote multi-stakeholder dialogues on AI innovation to enhance trust and adoption rates; 6) encourage initiatives on enhancing security and transparency; 7) promote the use of AI in small-and medium-sized enterprises; 8) increase training for human labor; 9) increase investment in AI; 10) encourage initiatives aimed at improving digital security and developing codes of conduct; 11) protect development privacy and data protection; 12) provide a market environment for the free flow of data.

UNESCO and COMEST issued the "Report on Robot Ethics" in August 2016, striving to promote "robots to respect the ethical norms of human society and write specific ethical norms into robot programing." It also proposed that the behavior and decision-making process of robots should be under supervision throughout the process.

The future global governance enabled by AI should not be determined by only a few countries or a few super companies. The generation and formulation process of AI rules, policies, and laws should incorporate multiple actors. In addition to international organizations, industry organizations, as social groups that take into account the functions of service, communication, self-discipline, and coordination, are pioneers and active practitioners in coordinating AI governance and formulating AI industry standards.

Industry organizations are composed of industry associations, standardization organizations, and industrial alliances. Representative industry associations include the Institute of Electrical and Electronics Engineers (IEEE), Association for Computing Machinery (ACM), Association for the Advancement of Artificial Intelligence (AAAI), etc. Standardization organizations include the International Organization for Standardization (ISO), International Electrotechnical Commission (IEC), etc. Industrial alliances include International Telecommunication Union (ITU), China's AI Technology Innovation Strategic Alliance (AITISA), AI Industry Alliance

(AIIA), etc. In order to promote all parties in the industry to implement the requirements of AI governance, industry organizations have carried out research that actively formulated AI technology and product standards.

In 2019, the IEEE released the official version of "Codes for AI Ethics Design," which advocated human rights, well-being, data autonomy, effectiveness, transparency, accountability, awareness abuse, ability, and other elements through an ethical research and design methodology.

Enterprises play a decisive role in promoting the implementation of AI governance rules and standards, and they are the backbone of carrying out rules and standards. As the main developer and owners of AI technology, enterprises have mastered a large number of resources such as capital, technology, talent, markets, and policy support. Therefore, enterprises should bear social responsibilities, strictly abide by scientific and technological ethics, technical standards and regulations, and implement self-restriction and supervision with high standards, so as to realize effective industry self-discipline and autonomy. When facing the social concerns caused by AI, some industry giants (i.e., IBM, Microsoft, Google, Amazon) have actively taken measures to ensure that AI can benefit mankind. These high-tech companies also put forward initiatives to ensure responsible design and effective testing of potential harmful biases in AI system, responsible use of AI and data, and reduction of injustices and other potential hazards in machine decision-making. In other words, enterprises are not only essential practitioners of AI governance rules but also important supervisors of AI moral development.

• International Cooperation Makes AI Go Far

Since national strategy is closely related with foreign policy, domestic politics, and international politics, plus the internal requirements for governance not only involve current global issues but also future risks, international cooperation is indispensable for global AI governance whether in international or industry organizations.

While AI technology is a scientific issue, AI governance focuses more on value construction that demands common understanding, cooperation, and norms. A government-led global cooperation network that involves non-governmental organizations needs to be established. This will enable the realization of optimal experience exchanges and effective responses to unemployment, the poverty gap, intelligent crime, and possible war threats. It can also help to build consultative and rescue platforms for humanitarian crises and anti-terrorism purposes.

The international community can properly respond to the new global governance needs and maximize AI service for the well-being of human society only through the linkage of AI strategic planning, transformation of government functions, enterprise innovation, ethical values construction, safety assessment, international cooperation and dialogue, and interdisciplinary comprehensive talent training.

On the one hand, flexible cooperation arrangements and binding commitments are mandatory due to the incomparable significance of information resources to state actors. In addition, technological innovation requirements may lead to a new power imbalance. In view of this, it is necessary to build a new international organization aimed at promoting AI governance and a more demanding international cooperation framework through flexible cooperation arrangements.

On the other hand, an international security cooperation mechanism is needed. When AI technology and data mining are applied for military purposes, it is intertwined with national sovereign interests and related to the control of future wars and international conflicts. Therefore, the military security application of AI is imperative.

In the long run, a strong collective supervision and enforcement mechanism may also help to curb national speculation and unilateral impulses in AI militarization (international conflicts may occur once countries insist on starting an AI arms race). At the same time, the AI governance mentioned above, especially government policies will also have obvious externalities and affect other countries. Therefore, the establishment of highly institutionalized and organized binding legislation, dispute settlement mechanisms, and law enforcement authority may help to promote global AI governance.

Again, the general path of global AI governance should take into consideration the construction of new legislation and rules. It should continue to promote AI R&D, address national security risks, ensure the public's acceptance of AI, and promote the good application and benign development of AI systems through international dialogue, coordination, and cooperation to enhance human welfare.

Basically, a timely establishment of AI ethical governance standards will help the technology go far, as it supports human society's moves towards a future of uncertainties. The prospect of AI is not entirely determined by technology per se; what the technology is used for is also a crucial question to ask in the face of the rising era of AI.

CHAPTER 8

China vs. US: Cooperation for a Win-Win Situation

8.1 The Third Thirty-Years in China-US Relations

The China-US relationship is one of the most complex and important bilateral relationships in the modern international environment. Since the Cold War, the US has become the only superpower in the world. But today, its global hegemony is being challenged by the rise of countries such as China. As China's global leadership power increases, the US finds her own power in this position declining.

On February 28, 2019, the famous Gallup Company released a poll on the global leadership powers of the US, China, Germany, and Russia in the previous year. The poll involved respondents from 133 countries around the world. Results showed that Germany ranked first with 40% support, which was the first time that Germany has fallen below this rate within the recent decade; China ranked second with 34% support, the highest in a decade; The US ranked third with 31% support, which was one percentage point higher than the previous year; Russia ranked fourth with slightly less than 30% support, almost equal to that of the US. The report concluded that "China's leading position has gained greater advantages in the competition among big countries."

Since the Reform and Opening-up, China has experienced forty years of rapid development and has become the second largest economy after the US. With the trend of China catching up with and surpassing the US becoming more and more obvious, China's sustained and steady development has aroused a deep concern in the US government. It is commonly agreed that China is the most likely challenger to the hegemonic status of the US, and the international community generally believes that China and the US form a challengers-hegemony relationship in the international system.

We live in a globalized, integrated, and diverse world of great power contests, where comprehensive national strength, science, and technology are above everything. The complexity and importance of China-US relations add more possibilities and uncertainties to AI development and governance.

• The First and Second Thirty-years

The first thirty-years of China-US relations refers to the Cold War period where China and the US were in a fierce confrontation.

In October 1950, four months after the Korean War broke out, China chose to enter the war when the US army crossed the Military Demarcation Line and frequently sent airplanes into China's northeast airspace. 2.4 million volunteers entered North Korea and fought bloody battles with 16 countries' armies led by the US, forcing the US to sign an armistice agreement in 1953.

Then there was the Vietnam War. Between 1964 and 1973, the US successively invested more than 540,000 troops to this war. More than 58,000 of them were killed and over 100,000 injured. China also sent more than 300,000 troops to Vietnam for the war. However, unlike in the Korean War, China did not directly participate but mainly helped Vietnam train its army, provide logistics, and formulate combat plans.

The hostility between China and the US eased in 1972 with President Nixon's visit to China. This was a far-reaching diplomatic revolution. After the establishment of diplomatic relations between China and the US and Chairman Deng Xiaoping's visit to the US in 1979, the cooperation between the two countries ended the cold war in East Asia.

The second thirty-years took place after China's reform and opening-up that led to continuous integration in the world economy. On January 1, 1979, China and the US established diplomatic relations, and Deng visited the US later that month. Since then, the two countries entered a period that focused on win-win cooperation. In 1982, China and the US jointly issued the "August 17th Communiqué" after consultations. Taking this opportunity, the bilateral relations entered a good state that was later widely known as the honeymoon period between China and the US.

During this period, China and the US maintained good interactions based on cooperation. According to the Gallup poll, at the beginning of 1989, 70% of Americans had a good or very good impression of China.

The international situation of this time was changing more prominently from bipolar to multipolar. In 1969, President Nixon introduced the theory of "five forces" in the international system, which means that the US, the Soviet Union, Europe, China, and Japan are gradually becoming the leading powers that shape the world. After China became one of the permanent members of the United Nations Security Council, China's international status and role could no longer be ignored.

Meanwhile, China walked out of the diplomatic disorder of the Cultural Revolution. The domestic situation was stable, and the national economy began to soar. With her independent foreign policy, China actively engaged in the international community and cultivated new China-US relations, introduced advanced capitalist technology and management experience, and vigorously developed productive forces. In 2001, China joined the WTO and achieved explosive economic growth.

There have also been some crises during this second stage of China-US relations, but both sides were willing to initiate crisis control, and the heads of state and senior officials of both countries could manage and resolve the tensions in a relatively short period of time, thus forming the situation of "fighting without breaking." Generally speaking, in the second thirty-years, China was continuously integrated into the US-dominant economic order and achieved greater global power and higher international status.

• Fighting Without Breaking: Confrontation Upgraded

After the 2008 financial crisis, China-US relations entered an adjustment stage and the third thirty-year period. In it, especially since Donald J. Trump's coming to power, the US has regarded China as an opponent with increasingly rapid and fierce suppression.

Although the US had also briefly regarded China as a competitor during the second thirty-year period, China wasn't considered as a dangerous opponent during the administration of George W. Bush (January 2001–January 2009) due to its general weak economic capabilities, despite rapid growth. In 2000, China's GDP was only $1,211.346 billion, which was equivalent to 11.78% of the US GDP at $10,287.479 billion. In 2008, China's GDP rose to $4,600,589 billion, which was 31.26% of that of the US at $14,718,582 billion.

By 2016 before Trump took office, China's GDP had reached $11,194.752 billion, which was 60.11% of that of the US at $1,862,447.5 billion. This aroused great vigilance and concern in the US. At this rate of development, by around the year 2030, China's total GDP will surpass that of the US and become the first in the world. This gives the US full motivation to suppress China.

Since the end of World War II, the opening of the US final customer goods market has become the source of its economic power. Similarly, the success of the East Asian model is also based on such an international economic division structure to a large extent. However, in the highly closed fields such as finance, currency, military, and technology, the US will take comprehensive suppression measures against its competitors once it believes they are causing a threat. For example, in the 1980s, the economic war between Japan and the US happened not only because of the increase of Japan's total GDP but also because Japan showed a strong potential in catching up with the US in some technical fields like semiconductors.

Therefore, the domestic attitude towards China changed widely in the late Obama Administration. After Trump took office, he reinforced it with response measures step by step.

In the 2017 "National Security Strategy Report," China was positioned as the US' primary strategic competitor, and the subsequent Indo-Pacific Strategy finalized the "road map" to contain and suppress China in various fields. On August 13, 2018, President Trump signed the "2019 Defense Authorization Act" that redefined Russia and China as "strategic competitors" and proposed to formulate a "whole government strategy for China."

In recent years, although the number of Chinese students studying in the US is still growing, this growth has slowed down year by year since 2015. According to public figures, the total number of students from mainland China studying in the US in the 2017–2018 academic year was 363,341, representing a yearly increase of 3.6%. In the 2018–2019 academic year, the total number of Chinese mainland students studying in American universities was 369,548, representing an yearly increase of only 1.7%.

Study abroad and scholarly visits have also become less certain with a twice or three times higher visa declination rate. Personnel exchanges represent the communications of not only knowledge and technology but also cultures at an individual level. Cooperation and personnel exchanges between enterprises should not be turned into a topic of national security, nor should they be blocked in the name of this reason. Now, under the context of "decoupling," the huge trade and frequent personnel exchanges between China and the US have changed from the link of bilateral relations to the content of strategic competition.

In addition, the escalating trade war and technology war have undermined economic and trade relations and pushed China-US relations on thin ice. After two years of escalating tensions, China and the US signed the "Economic and Trade Agreement between the Government of the People's Republic of China and the Government of the USA" in January 2020 in Washington, D.C. But before this agreement could bring relief to the China-US relations, the outbreak of the COVID-19 pandemic made it further deteriorate.

Fighting this large-scale infectious disease that has rapidly spread all over the world and endangered human beings as a whole failed to ease the tensions between China and the US.

Friction and contradictions constantly arose between the two countries. The source of the virus, stigmatization, and response mode were the main disputes between China and the US.

The outbreak of the coronavirus pandemic in early 2020 provided new energy for the Trump Administration to suppress and disconnect with China. It widened the China-US differences in reality, exacerbated the mutual political suspicions, and deepened public hatred against each other.

The pandemic began to spread rapidly in the US after March 2020. The controversy between China and the US on the source of the virus escalated. Secretary of State Michael Pompeo acknowledged China's medical supplies but also publicly stated that COVID-19 originated from China. On April 21, Missouri became the first state in the US to seek an indictment for the way China responded to the pandemic. Although it is very difficult to require China to be responsible for this litigation and the possibility of obtaining financial compensation is very low, China's assets in the US may face risks.

The deteriorating tension between China and the US brings up the question of how great powers should compete peacefully in the modern world. The poisonous atmosphere between China and the US has undoubtedly hindered bilateral cooperation and also weakened the confidence of other countries. More attention should be paid to the voice of seeking common ground while reserving differences.

At this point, it is worth pointing out that integrated cooperation and competition is still the main characteristic of China-US relations, where cooperation is greater than competition.

Given the huge cost of great power confrontation and conflict and the challenges of common global issues, the win-win doctrine should be the common value system and common cultural belief of future bilateral relations that form an unbreakable community of interests and responsibilities that benefit both sides.

8.2 AI Competition between China and the US

Scientific and technological capability is not only the core symbol of comprehensive national strength but also an important embodiment of national development potential. Right now, AI has become the focus of China-US competition.

AI the transformative technology has great potential to promote industrial innovation, improve economic benefits, and promote social development. Because of its capability to lead technological development and social transformation, AI is not only regarded as the "technological foundation" of the future innovation paradigm but also commonly regarded as the key force to promote new technological revolution and industrial change.

Throughout history, the couplings and interactions of the technological revolution, the industrial revolution, and military reform can deeply affect and even reshape global competition patterns. Therefore, in the global game of AI, the preemption of AI high ground by China and the US is crucial to the reconstruction of the future international structure and global AI governance.

- **US in the Lead; China Catching Up**

In 2019, the data innovation center of the Information Technology and Innovation Foundation (ITIF) released a hundred-page research report *Who Will Win the AI Competition: China, the European Union, or the US?* to compare and calculate the development status of AI in the three areas. The US took the lead with 44.2 points, China ranked second with 32.3 points, and the European Union ranked third with 23.5 points.

The reason that the US can occupy the leading position is closely related to the history of national AI development. In 1956, AI was officially born in the US. Carnegie Mellon University, MIT, and IBM became the first three core AI research institutions.

From 1960s to early 1990s, great progress was made in language programming, expert systems, and commodification. For example, in 1983, the world's first mass production company of a unified computer was established. Moreover, the AI research results were applied in practice. For example, the first mineral deposit in the US State of Washington was discovered using the mineral exploration expert system PROSPECTOR.

In contemporary China, AI has just entered the embryonic stage. In 1978, the China Science Conference was held in Beijing, and it provided the foundation for the development of China's AI industry. In the same year, "intelligent simulation" was incorporated into the national research plan, and China's AI industry began to officially develop with governmental support.

In terms of research results, the US is undoubtedly in a leading position in the world. According to the search results of Scopus, the world's largest citation database, 16,233 peer-reviewed papers on AI were published in the US in 2018. The rapid growth in the number of papers mainly occurred in 2013. Although the number of AI papers in China and the European Union also increased at this time, and the overall publication has significantly exceeded that in the US, the quality of US papers was significantly higher. In 2018, the average time for each US paper to be cited was 2.23, while that for a Chinese paper was 1.36. The number of citations per author in the US was also 40% higher than the global average.

In the field of Deep Learning in particular, the number of papers published in the US far exceeds that of other countries. From 2015 to 2018, 3,078 relevant papers were published on the preprint library website arXiv, which was about twice that of in contemporary China. In recent years, the number of AI patents obtained by the US each year account for about half of the global total, and the number of patent citations accounts for 60% of all citations in the world.

In terms of key technologies, the research results of the US are also in the lead. In the field of computer vision, the Noisy Student methodology developed by Google and Carnegie Mellon University reached a Top-1 accuracy of 88.4%, an increase of 35% over six years before. The time required to train a large-scale image classification system on the Cloud infrastructure has been reduced from three hours in 2017 to 88 seconds in 2019, and the training cost has decreased from $1,112 to $12.6.

In terms of industrial development, according to the "Research Report of Global AI Industry Data (2019 Q1)" released by the *China Academy of Information and Communications Technology* (*CAICT*), by the end of March 2019, there were 5,386 active AI enterprises in the world. Among them, 2,169 were in the US. 1,189 were in mainland China and 404 were in the United Kingdom.

In terms of historical statistics of enterprises, the development of AI enterprises in the US started five years earlier than in China. Emerging in 1991, the US AI enterprises entered the major development stage in 1998 and had a period of rapid growth in 2005. This development has gained stability since 2013. Chinese AI enterprises emerged in 1996 and entered a development stage in 2003. The industry gained stability after reaching its zenith in 2015.

US companies emphasize patents and leading AI acquisitions. Among the 15 machine learning subcategories, Microsoft and IBM applied for more patents than any other entity in eight subcategories, including supervised learning and reinforcement learning. US companies are leading patent applications in 12 out of 20 fields, including agriculture (John Deere), security (IBM), and personal equipment, computers and human-computer interaction (Microsoft).

Talent reserve is another key reason for the US to lead in the AI industry. In a way, the competition of AI is the competition of talent and knowledge reserves. Only by investing more researchers and continuously strengthening basic research can more intelligent technologies be obtained.

Apparently, American researchers pay more attention to basic research, which enables a solid foundation and significant advantage for the US AI talent training system. Specifically, the US has formed a pattern that can lead the world in key links such as basic discipline construction, patent and paper publication, high-end R&D talent, venture capital, and leading enterprises for a long time.

According to the research of MacroPolo think tank, 59% of the top AI research talent works in the US, 11% in China, and the remaining distributed in Europe and Canada. The talent differences are obvious.

• To Go After or to Go Beyond?

Although the US has the first-mover advantage in research results and talent reserve, China, as a rising power in AI, is catching up under supportive policies and an open application environment.

After comparing government preparation for AI of different countries, Oxford Insights ranked the US government fourth in the world after Singapore, the UK, and Germany. In addition, the US ranked top in key indicators such as innovation ability, data availability, AI standards in the government, labor skills, number of start-ups, digital public services, and government effectiveness.

China came in 20th in the ranking. It was believed that China's biggest weakness was the backwardness of basic research while its comparative advantages were government's emphasis on high-tech development, rich data, loose supervision, as well as the rapid growth in the number of engineers. After years of accumulation, China has made a series of important achievements and formed its own unique development competencies. Speedy development is seen in top-level design, R&D resource investment, and industrial development, and some core technologies can also compete against that of the US.

The Congressional Research Service (CRS) unmistakably expressed the concern of China being the most insidious competitor of the US in 2019: "Potential international competitors in the AI market are creating pressure on the US in innovative applications of Military AI ... So far, China is the most ambitious competitor of the US in the International AI market."

PricewaterhouseCoopers also reported that in the era of AI, no other country can catch up with the US or China in terms of technological development or national strength; and neither the US nor China can monopolize this field or coerce each other. By 2030, the US and China will account for 70% of the $15.7 trillion wealth brought by AI to the global economy. The unique advantages of the two countries include first-class professional research knowledge, deep capital pools, rich data, supportive policy environments, and highly competitive innovation ecosystems. At present, about half of the global AI companies are in the US and 1/3 are in China.

Both countries emphasize top-level design. In October 2016, the Obama administration released two important documents on AI development: the "National Strategic Plan for AI R&D" and "Preparing for Future AI." In March 2017, the Chinese government included AI in the national government work report for the first time and released the "Development Plan of New Generation AI" in July the same year, which made AI a national strategy in an all-round way. In other words, Both China and the US have established a relatively complete R&D promotion mechanism at the national level to promote the development of AI as a whole.

In comparison, the US government's investment in R&D is insufficient. Over the past few decades, the percentage of federal spending on R&D in GDP has decreased from 1.86% in 1964 to 0.7% in 2018.

Right now, the annual fiscal deficit of the US Federal Government has exceeded $1 trillion, and the accumulated government debt is equivalent to 107% of GDP. These factors will limit the long-term government investment in AI and related basic research.

However, in China and the EU, government investment in R&D is increasing and gradually catching up with US funding. Between 1960 and 2016, the share of the US in global R&D investment decreased from 69% to 28%. From 2000 to 2015, the US accounted for only 19% of the global R&D investment growth while China accounted for 31%. On August 31, 2019, Shanghai announced the establishment of the AI Industry Investment Fund. The first phase investment alone was CN¥10 billion, and the final scale will reach CN¥ 100 billion.

Viewed from the perspective of industrial development, although the comprehensive strength of the basic layer of AI industry in China is weak and that there are few global leading chip companies, major manufacturers such as Baidu, Alibaba, Tencent, and Huawei are accelerating their laying out speed to catch up with the US standard.

At the technical level, Chinese enterprises have a good momentum of development. Comprehensive manufacturers have the layout in multiple core technology fields such as computer vision, natural language processing, and speech recognition. Meanwhile, entrepreneurial unicorns are developing rapidly in the vertical field.

In the application layer, Chinese AI enterprises have achieved extensive layouts in education, medical treatment, new retailing, and many others. A large number of AI enterprises are also competing in finance, medical treatment, retail, security, education, and robotics.

Compared with the US, China has two important advantages in AI development.

First, the management system of China's AI ecosystem is very different from that of the US, which features a "strong market" and "weak government." The past technological revolutions in the US were mostly dominated by scientific and technological enterprises or institutions, and the role of the government in industrial development has always been limited. For China, the government plays an important role in the economy. With the strong support of national policies, the top-down transformations are easier to achieve. By strengthening the cooperation and resource sharing of the whole society, the government may fully occupy the commanding heights of information technology and become a leader in AI.

Second, China's data collection barriers and data labeling costs are lower, making it easier to create large databases, which is an essential basis for the operation of AI systems. By 2020, China is expected to have a global data share of 20%; By 2030, this share may exceed 30%.

In a way, China also has an important advantage in big data. China's relatively complete industrial system and huge population base make China's AI development advantageous in data accumulation.

While it is difficult to predict what the future of AI may look like, it is certain that only through mastering the strategic cutting-edge technologies and becoming the rule-maker in the new competition paradigms can a country truly win in the future.

8.3 Unite in a Concerted Effort

As China quickly is growing into a great power after the US in the international system, strategic competition between China and the US is inevitable. China-US relations are both influenced by the law of competition among great powers and restricted by the development trend of the times. As the most complex and important bilateral relationship in the world, the ups and downs of China-US relations are increasingly affecting international orders and global situations.

The world is moving in a direction of unprecedented advancement and unpredictability led by new technologies such as AI. On the one hand, as Russian President Vladimir Putin said, AI determines a country's future. Whoever masters it can stand out in international competition and obtain huge competitive advantage. On the other hand, AI also brings non-traditional security challenges to global governance. It intensifies some current crises and also brings up new problems. This derives not only from technological issues but also from a series of problems brought by the combination of AI and international politics.

In this context, competition and cooperation will become the main path for the development of China-US relations in the future, whether in theory or practice, history or reality, domestic or international, combat or peace. How to grasp and make good use of opportunities and how to obtain greater benefits from cooperation have become a strategic plan for the development of China-US relations.

- **Conflict deepened by AI**

Two scholars of the Brookings Institution believe that military conflicts have the greatest risk of miscalculation. In consideration of technical security, countries generally take a more cautious attitude towards the militarized application of AI. The international community also calls for limiting AI applications in this field and preventing the proliferation and irresponsible use of AI weapons. What's more, the militarized application of AI is not confined to a single type of weapon or combat platform. Therefore, the party with technical advantages will obtain absolute advantages in strategic judgment, strategy selection, and execution efficiency; while the party with technical disadvantages has difficulty making up for this gap with other means such as quantity stack or strategy and tactics.

Based on this understanding, every single action taken by either side in the China-US relations will make the other side feel insecure and stimulate countermeasures. As AI technology is increasingly integrated into weapon systems, this security dilemma may become more obvious, leading both China and the US to nationalize innovation with lower transparency in order to seek advantages over each other.

Viewed from the perspective of the developmental contradiction of science and technology, the global industrial division of labor and technological exchanges promoted by multinational enterprises have reached a high level. This is one of the important foundations for the rise of AI upsurge. Take the industrial division and cooperation between Chinese and the US enterprises as an example. Among the names of core suppliers (92 in total) released by Huawei in 2018, although some enterprises suspended their supply to Huawei due to the impact of the US ban, 33 US suppliers still made the list and accounted for the largest percentage. Among the top 200 suppliers of Apple in 2018, China's enterprises accounted for 47.6%. (41 are in mainland China).

The suppressive measures that the US took against other countries can easily induce a "technology cold war" and erect a new "Berlin Wall" in science. That is, when countries adopt protective actions in the field of AI and related technologies, international cooperation that promotes technology development (i.e., technology, investment, talent flow) may even face the risk of being completely restricted.

Viewed from the perspective of geopolitical conflict, AI is inevitably related to geo-economic competition and political problems as one of the important technologies in the competition among great powers. It is generally believed in the US that AI technology will become a tool to intensify ideological competition between China and the US, especially when one or both parties use this technology to interfere with the other's domestic political affairs, such as in 2016 when some people firmly believed that Russia had intervened in the US presidential election.

In addition, different countries will have different development modes based on national conditions, including developmental stages and demands. China, the US, the European Union, and other parties have great divergence of opinions in data acquisition, use, and processing of face recognition. In comparison, the European Union most vigorously emphasizes the protection of personal digital privacy, which is also highly stressed in the US. But China has relatively loose supervision on face recognition technology. The different development models and interest demands often lead to greater division and more severe competition in the technology industry.

• The Possibility and the Inevitability of Cooperation

In an age where "great changes unseen centuries" take place, AI as the core technology of a new technological and industrial revolution is of great importance to global development. However, the uncertainty of AI development and the instability of its application have also exacerbated the difficulty of technology risk management and control, bringing new challenges to global governance.

First, high productivity means high wealth. Jack Ma (Ma Yun) believes that "AI will completely change the mode of human employment." It is thus obviously more advantageous to countries with higher innovative ability and an entrepreneurial economy. Aside from replacing human labor and creating huge material wealth, AI can also promote a country's comprehensive social governance level, economic innovation and development efficiency, national defense and security construction, etc. This means that there will be a wider distinction in the ways and speed of wealth accumulation and in international strength among countries. The rich and strong will get richer and stronger, while the poor and weak will get poorer and weaker. More wealth inequality, injustice, resistance, conflicts, and terrorism will arise, and greater uncertainty and difficulties in global governance will occur.

Second, the high executive power means high destructive power. Emerging new weapons and cyber viruses have the ability to respond quickly and act tirelessly—something that humans cannot achieve. While such capabilities can be beneficial in implementation, they are vulnerable to forces that can cause serious security hazards to the international community and even disasters to the entirety of human society. Although corresponding solutions may be found when there is a serious crisis, the harm or potential harm will nonetheless consume great numbers of human, material, and financial resources and significantly increase the costs of global governance.

Third, high intelligence means high politics. Advanced AI gives technology owners outstanding additional advantages. The vigorous development of AI in areas such as the robotics industry, gene sequencing, automatic driving, intelligent finance, smart cities, big data processing, natural language processing, image recognition, and intelligent military systems has changed the core competitiveness and national economic, social, and industrial structure—meaning the power structure—in many ways.

Power structure is a bargaining chip in political deals. Constantly consolidating and pursuing power is the core motivation of national foreign strategy and policy. AI will have an important impact on international relations from the field level, institutional level, and ideological level. These influences will directly affect the subject and object of global governance, thus affecting the effectiveness of global governance.

As great powers in the development and application of AI technology, China and the US have unique advantages that cannot be duplicated. Therefore, to prevent or to reduce the negative impact of AI technology progress on global sustainable development and strategic stability will require the full cooperation of the two countries.

In fact, the competition in the field of AI is not an absolute zero-sum game. Both parties can benefit from cooperation. While China is leading more in experimental research and application of achievements, the US is in the leading position in basic research and frontier technology exploration. Of course, it is not surprising that there should be differences in the principled positions, interest demands, and policy propositions of the two sides. The key lies in how to rationally view such differences and effectively control the potential conflicts.

Therefore, China and the US should work together to launch a formal discourse on AI application in security and economy, promote the transparency of AI research and development, promote the rational distribution of its beneficial achievements worldwide, avoid great power competitions that may lead to catastrophic conflicts to the greatest extent, and strive to form a rational and benign cooperative relationship.

As Henry Kissinger said, China and the US are two countries that are most capable of influencing world progress and peace in terms of technology, political experience, and history. It will be the common global responsibility of both countries to solve important matters in a cooperative manner.

8.4 From Nuclear Arms Race to AI Cooperation

After World War II, on the basis of deep reflection on conflict and war, an international dispute settlement mechanism with the United Nations as the center was established internationally. Important professional international organizations such as the International Monetary Fund, the World Bank, and the WTO were also founded. At the same time, with the booming of national independence movements, the democratization of international relations, the regularization of international order, and the transparency of international politics became the important topics of the time. The relationships between major countries with better cooperation rather than stronger competition is gaining wider recognition.

In his article "Remember the history and create the future," published in *Rossiyskaya Gazeta* (a Russian newspaper), President Xi Jinping pointed out that "Peace rather than war, cooperation rather than confrontation, win-win rather than zero-sum" is the eternal theme of peace, progress, and development of human society. In facing the new global anxiety caused by AI, China and the US, as AI great powers should follow the trend of "cooperating through competition, and competing through cooperation," maintaining the understanding that cooperation is greater than competition, and cooperate towards a win-win situation.

• Competition and Cooperation in the Era of the Nuclear Race

Great power relations can have different manifestations in different periods. Cooperation and competition are two sides of the same coin. Great powers can both be allies and enemies.

During World War II, the Yalta System was formed, centering on the agreements and ideas of the Yalta Conference. The system aimed at the coordination, cooperation, and co-governance of major countries, and it established the basic framework of post-war international order and global governance. The US and other western countries played an important role in this process. As the global governance mechanism of the international community for world peace and security, the Yalta System made arrangements in line with the international development trend for the post-war international order, which effectively restrained the fierce competition among big countries in the early stage of its establishment.

However, the aftermath of the war had a long-lasting impact on the post-war international order. The irreconcilable power contradiction, interest contradiction, and hostile ideology between the US and the Soviet Union still highlighted the existence of great power competition. For forty five year, between the formation of the Yalta System in 1945 and its disintegration in 1989, the US-USSR relations has always been the main factor that shaped the postwar world patterns.

The period between the late 1940s and the mid-1980s witnessed the alternation of tension and mitigation between the US and the Soviet Union under the Yalta System. During this period, the growth and decline of the two superpowers had a far-reaching impact on the changing of world patterns. From the late 1940s to the late 1960s, the Cold War had the main manifestation of a bipolar pattern. The US mainly took the strategic offensive position with some level of defense, and the Soviet Union was in a strategic defensive position with some level of attack.

The nuclear arms race showed that competition outweighed cooperation in the US-USSR relations. During the Cold War, the US and the Soviet Union launched a fierce contest around the competition for nuclear advantage. The pursuit of explosion equivalent, accuracy, and miniaturization of nuclear weapons pushed the power of nuclear weapons to extremity.

At the same time, in order to avoid direct collisions, to ensure the "stability" of this contest without endangering respective safeties, and to maintain the nuclear advantage and monopoly position of both sides relative to other countries, the US and the Soviet Union also conducted a series of arms control negotiations. The interaction with an aim for nuclear superiority and nuclear arms control for the purpose of ensuring the advantageous position of both sides led to the growth and decline of their nuclear arsenals.

On July 16, 1945, the US successfully exploded its first atomic bomb, making it the first country to master and gain a monopoly in nuclear weapons. Facing the grim threat from the US, the Soviet Union sped up in its own nuclear weapon development and broke the US atomic monopoly in August 1949. In addition, it took the lead in mastering thermonuclear weapon technology, and a situation of mutual deterrence between the two countries was formed.

The fear of the complete destructive power gave rise to the voice of "mitigation" in both countries. After Nikita Khrushchev came to power, he put forward a "peaceful coexistence, peaceful competition, peaceful transition" plan, and the contemporary Eisenhower Administration put forward the strategy of "peaceful victory" that advocated improving the relationship with the Soviet Union through dialogue and negotiation. The following US and USSR leaders John Kennedy and Leonid Brezhnev had both taken the element of "mitigation" into consideration in their policy making.

In December 1953, President Eisenhower delivered the famous speech "Atoms for Peace" at the United Nations General Assembly. He proposed to establish the International Atomic Energy Agency (IAEA) that required all countries to donate nuclear materials and to use nuclear energy for peaceful purposes (i.e., agriculture, medical treatment, power generation, etc.) The US, the Soviet Union, and other countries producing nuclear materials have conducted long-term negotiations on the statutes of the IAEA. The central argument was on the agency's control over nuclear materials under its possession and whether the agency has the power to inspect bilateral or multilateral agreements on non-military materials. A final consensus was made in October 1956 and the "Statute of the IAEA" was passed. It came into force the following year, and the IAEA was officially founded.

The IAEA is the initial institutional arrangement for global governance of nuclear security. According to the Statute, the IAEA has three guiding ideologies: 1) safeguards and verification—to verify that nuclear materials and activities of states are only used for peaceful purposes in accordance with legal agreements and carry out safeguards inspections; 2) safety and security—to formulate safety standards, norms, and guidelines to help member states throughout the application process; 3) science and technology—to provide technical and research support for nuclear applications in health, agriculture, energy, environment, and other fields.

In 1962, the outbreak of the Cuban Missile Crisis shocked the world. The Soviet Union deployed missiles in Cuba after the US deployed the medium-range ballistic missiles Thor IRBM and PGM-19 Jupiter in the UK and Turkey in 1959. The Cuban Missile Crisis was the most intense confrontation between the Soviet Union and the US during the Cold War. Although it lasted only 13 days, the two countries' hovered fingers over the nuclear button have never brought mankind closer to the edge of destruction.

The Cuban Missile Crisis made both the US and the Soviet Union deeply realize that avoiding nuclear war should be the absolute top priority at all times, and it is wise to accept the existence of a hostile government. The Cuban Missile Crisis was also one of the factors that gave birth to the nuclear test ban agreement. After the crisis was resolved, the US and the Soviet Union built a hotline connection in December 1962. Kennedy and Khrushchev kept exchanging letters to speed up the negotiations on banning nuclear tests.

Soon after, the US, the Soviet Union, and the United Kingdom resumed negotiations in Moscow. On August 5, 1963, the representatives of the three countries signed the "Treaty of Banning Nuclear Tests in the Atmosphere, Outer Space, and Underwater," that is, the "Partial

Nuclear Test Ban Treaty." It was of some significance in reducing nuclear pollution of mankind's environment and easing the confrontation between the US and the Soviet Union.

However, the US government's easing policy soon received much domestic opposition. With the rising conflicts in Angola and the Middle East, the easing atmosphere of bilateral relations deteriorated. In 1979, the Soviet invasion of Afghanistan and President Reagan's hard-line policy towards the Soviet Union intensified the bilateral relations again and escalated the arms race.

Mikhail Gorbachev came to power in the mid-1980s. Under the guidance of "new political thinking," he not only made major adjustments to domestic policies but also to the view on war and strategic thought. Gorbachev put forward the principle of "sufficient defense," which refers to the idea of not feeling compelled to compete against every single military progress that the US has made under the condition of having "reasonable and sufficient" amounts of nuclear power. Therefore, although the Soviet Union continued to develop its nuclear forces, investment was no longer made in deploying new types of weapons, and the Soviet Union's nuclear forces were basically finalized.

In addition, Gorbachev formulated many nuclear arms control proposals and accepted the "Intermediate-Range Nuclear Forces Treaty," which eliminated medium range nuclear forces in an unbalanced manner between the US and the Soviet Union. As a response to the adjustment and reform taken by the Soviet Union, President Reagan resumed dialogue and negotiations with the Soviet Union in the later period of his term of office.

The large number of strategic weapons possessed by the US and the Soviet Union were important in maintaining the balance between the two superpowers during the Cold War. The cost of nuclear war is far beyond people's imagination. It will not only cause extreme casualties on both sides for combatants but also for regular citizens. Neither belligerent can win alone in a nuclear war.

Although in the time of bilateral confrontation, the Yalta System was seriously constrained by the tension between the United Nations and the Soviet Union and did not achieve its desired effect, there was still some cooperation going on between the two countries, which implied that the Yalta System still played a certain role in assuring world peace. The US-USSR relations have important practical significance for preventing regional conflicts and dealing with international security challenges in the era of AI.

• Competition and Cooperation in the Era of AI

There are things that "change" and things that "do not change" in international cooperation. The former refers to the change of mutual relations between countries over time and conditions, including adversarial relations, dependency relations, alliance relations, partnerships (competition and cooperation), as well as relations in the fields of economy and trade, science and technology, and security. The latter refers to the basic elements of cooperation between countries that remain unchanged, such as business and trade, personnel exchanges, medical and health care, science, technology and education, and common problems and challenges.

Compared with the era of nuclear competition, the role of great power games has changed. Between the US-USSR nuclear competition and the US-China AI competition, what have changed are strength, purpose, relationship forms, and allocation of power. What remain unchanged are the basic element of international cooperation, and the mutual respect and justice that enable a win-win situation. In the face of new challenges of international security and global governance, cooperation is the only solution for China and the US.

China-US coordinated soft governance

The most important way to promote global AI governance is to keep pace with the times and adjust the coordinated and common governance mechanism of the great powers. The Yalta System formed after World War II was a global governance mechanism dominated by Western countries. After the Cold War, it gradually lagged behind the needs of the development of global governance. Governance will be more legitimate and targeted only if emerging powers such as China, Russia, and India cooperate with old powers to jointly deal with the challenges of global issues.

The relations of states are essentially relations of people. The fundamental driving force and practical purpose of bilateral cooperation lies in benefiting people in both countries. In other words, individual interests need to be guaranteed by the state, and any global governance mechanism must be implemented at the national level. Based on the previous examples of technology-state relations, inappropriate handling of technological dominance competition can lead to wars.

In the Fourth Industrial Revolution, the US plays the role of rule maker and China appears as an emerging force. Therefore, the US endeavors to curb China as a major competitor, which has been obviously manifested in the China-US trade disputes in recent years. In essence, this trade dispute is a technological competition. The US wishes to prevent China from gaining the right to define in the new technological revolution. This is still looking at the current problems with the old Cold War thinking, which will inevitably lead to conflict.

To get out of this thinking, it is necessary for the two countries to combine their efforts under the new framework and jointly contribute to the scientific and technological progress of mankind. In fact, in the era of big data, the concentrated embodiment of Cold War thinking is data nationalism.

In the view of data nationalists, data will be safe only if it is stored in their own country. At present, this concept is increasingly adopted by the policy makers of various countries and integrated into national policies related to data. Data nationalism requires all countries to form a series of conventions on the application of AI in global governance, and China and the US should work together to promote the establishment and implementation of such a convention.

John Frank Weaver suggests in his book *Robots Are People Too* that a series of international conventions should be made on AI and national responsibility, national sovereignty, self-guided vehicles, intellectual property rights, monitoring, and armed conflict. Weaver points out that there should a multilateral agreement that determines how AI affects national sovereignty, what degree of UAV surveillance is allowed, and what AI can do in armed conflict.

This idea of multi-national convention emphasizes that coordination among countries should not only be based on international conventions with fixed rules and binding commitments; rather, there should be some flexible soft governance frameworks such as non-confrontational and non-punitive compliance mechanisms on a voluntary basis. For example, ASEAN's regional governance is famous for its inclusiveness, informality, pragmatism, convenience, consensus building, and non-confrontational negotiations, which is in sharp contrast to "hostile posture and legitimate decision-making procedures in Western multilateral negotiations." It represents a new form of global governance, similar to what is shown through the Paris Agreement on climate change. Therefore, multinational cooperation in the field of AI is also possible to be carried out in the form of soft governance in the future.

Promote Rational Distribution of AI Achievements

China and the US should reach a consensus on the development of general AI and promote the rational distribution of its beneficial achievements around the world.

On the one hand, China and the US have the responsibility to help achieve a balance between global and national governance through AI. In order to achieve such balance, it is necessary to form a coordination mechanism at the global level and form a positive interaction with domestic systems. At present, the international rules related to AI are mainly promoted and formulated by enterprises or institutions in western developed countries, which means that a "global" AI governance mechanism is still absent. Therefore, more developing countries are required to join the process. Additionally, global governance should eventually adapt to the framework of national governance. Nation states are still the most important actors in the international community, and important issues on personal welfare and social security still need to be solved by the state government. Therefore, the most critical question to ask for national governance is how to improve the national governance capacity of all countries (especially developing countries) through the technology and industrial development of AI, so as to fundamentally eliminate the domestic root causes of global problems.

On the other hand, while developing countries can have more opportunities to improve their national conditions through the huge enabling capacity of AI, this process requires China and the US to promote through rationally distributing the useful achievements of AI. Simply put, all that is needed is a more equitable way to allocate the future growth. In this light, intellectualization should be the optimal solution to most global problems. For example, many developing countries suffer from the shortage of water resources. This can be effectively solved through intelligent water resources management. Additionally, when it comes to the influence of social traditions on political management, intelligent systems and equipment can help minimize the negative impact of social traditions.

In sum, AI should be promoted as an important scheme to solve global problems. Since the biggest advantage of AI is to save human resources, and that the biggest challenge for the United Nations and other international organizations to carry out tasks in developing countries is the lack of human resources, AI can also play a complementary role in this regard.

Develop Interpretable AI

The path from weak AI to general AI calls the attention of all great powers. It is important to remember that AI cannot solve all problems once and for all, and the development of general AI may also challenge the significance of human existence.

Karl Marx pointed out in *Economic and Philosophic Manuscripts of 1844* that "The characteristic of mankind is free conscious activities." These "free conscious activities," or human subjectivity, needs to be confirmed and reflected in practical activities. Through productive labor, people have changed the existing form of nature and realized human ideas and goals so as to control nature. However, the problem of AI alienation has fundamentally challenged human subjectivity and practicality.

China and the US should thus promote the transparency of AI research and development, reach a consensus on the development of general AI, and confirm which types of general AI can be developed and which should be prohibited. It is thus necessary to fully evaluate the types of general AI and the overall development consequences, especially the development of higher-level general AI.

In other words, interpretable and safe AI is the future. AI scholar Amir Husain believes that the world is in urgent need of ultra-high safety and interpretability standards to promote the development of AI.

The current wave of AI is mainly driven by Deep Learning algorithms, which is a feature quantity automatically generated by computers based on data. DL does not need to be designed by people but can automatically obtain the high-level features with a computer. The typical DL model is a deep neural network. For the neural network model, a simple way to improve the capacity is to increase the number of hidden layers. With the increase of the number of hidden layers, the corresponding parameters such as neuron connection weight and threshold will increase together with the complexity of the model. DL emphasizes data abstraction, automatic learning of features, and attention to connectivity.

However, the theory of DL is still imperfect. Thus in the application practice of DL, engineers need to manually adjust parameters to get a better model. But at the same time, engineers cannot explain the influencing factors of the model effect. In other words, due to the un-interpretability of internal parameters, DL has an algorithm black box effect. Therefore, the development of interpretable AI becomes the key. A knowledge map, which applies methodologies such as entity chain index, relationship extraction, knowledge reasoning, and knowledge representation, is likely to promote the development and achieve a breakthrough in this regard.

8.5 Chinese Wisdom for AI Global Governance

By the year 2030, China will achieve the goals listed in *The Development Plan for a New Generation AI*—to reach the world-leading standards in AI theory, technology, and application and become the world's main AI innovation center. By then, the gap in economic strength between China and the US will be further narrowed, and the economic status of the two countries may even be reversed. In terms of military power, China is also rapidly catching up with the US.

As an important representative of developing countries and a responsible global member, China has the duty to promote the safe, peaceful, and just use of AI, manage the positive development of AI, and make sure the AI achievements benefit the majority of developing countries in the world.

- **Cooperate for Win-win**

From ancient times, the Chinese nation has been a nation that adores peace and common development.

"Harmony" and "peace" are important elements in Chinese culture and the Chinese humanistic spirit. It embodies the Chinese people's thinking about coexisting with nature, society, and mankind. It is the way of communication and the basis of cooperation.

Peaceful coexistence between China and the US is of great significance for promoting mutual cooperation, promoting their respective and common interests, and maintaining world peace, development, and stability. After the Cold War, "peace benefits both sides" are reflected in many aspects in the China-US relations, such as investment, trade, education, culture, and health care.

On the one hand, China has shown some level of support for the US in the Gulf War, the War on Terror, the War in Afghanistan, the War in Iraq, as well as on issues such as nuclear tests in North Korea, the Nuclear Program of Iran, and climate governance. On the other hand, the US imported a large number of goods from China, decoupled MFN treatment from human rights issues, supported China's accession to the WTO, and allowed China to play a bigger role in international organizations such as the United Nations and the International Monetary Fund. All of these measures have helped China to achieve vigorous development.

The win-win cooperative relationship between China and the US is determined by the development trend of the two countries in the new era of globalization. It is not only in line with the fundamental interests of the Chinese and American people but also productive to the smooth progress of the transformation of the international order and the reform of the global governance system. President Xi Jinping pointed out that "Achieving non-conflict, non-confrontation, mutual respect, and win-win cooperation between China and the US is the priority of China's foreign policy." Non-conflict and non-confrontation are the prerequisites, mutual respect is the foundation, and the win-win cooperation is the goal.

In order to achieve this goal, we need rational and fair competition and better cooperation based on consensus and rules. As the two most influential countries in the world, China and the US cannot effectively promote global governance without mutual cooperation. Only through coordination and cooperation of great powers can some major issues related to global interests be effectively solved.

• Adhere to Common Security and Promote Common Development

When it comes to AI security, "Universal security is the greatest security, and common security is the best security." Without sustained security, there will be no peaceful and sustainable development.

On the one hand, in view of the problems that the development of AI technology may lead to (i.e., international public security, extreme poverty gaps), the Chinese government should call on the international community to strengthen the monitoring of the AI developmental risks, actively submit proposals to relevant international organizations, advocate the establishment of technical management frameworks and ethical norms, cooperate with the international community to control development risks (similar to controlling the nuclear weapons), strictly control AI applications, and establish international conventions to avoid disastrous consequences for human societies.

Examples are:

1) Actively participate in international control negotiations on lethal autonomous weapon systems, control the development and deployment of AI weapons, ensure that key military decision-making powers should be controlled by humans and not machines, resolutely oppose completely autonomous weapon systems, limit AI weapon abuse, and maintain the stability of the international community;

2) Promote the establishment of an effective communication mechanism between countries to avoid misjudgments;
3) Promote the establishment of an international organization, similar to the IAEA, aimed at promoting cooperation, reliability, and peaceful use of AI technology, and assures that it has the power to supervise the development of AI weapons in member states;
4) Give the UN Security Council the power of enforcement;
5) Make commitments and set an example to establish its own moral position and principles.

On the other hand, China should also actively participate in international discussions on the technical standards and development principles of AI, to make sure that the development of AI is "people-centered" and in accordance with common ethical values. China should engage in developing beneficial AI and oppose malicious exploitation, lead and promote the formulation of unified global standards and formulate standards beneficial to China and the majority of developing countries.

Finally, as a representative of the developing countries, China should make full use of the Belt and Road Initiative (BRI) to provide technical, financial, and educational support to developing countries in adjusting to the era of AI and to reduce the "digital gap," "technology gap," and "poverty gap" in the international community. For example, enhance data circulation security, strengthen cyberspace governance, and expand the application of big data. In March 2018, China's high-tech enterprise CloudWalk Technology signed a strategic cooperation framework agreement with Zimbabwe and took the opportunity of the Belt and Road Initiative for China's AI industry and technology to enter Africa.

In the new era, China's participation in global AI governance should not only strive to seize the commanding heights of high-tech R&D but also pay attention to gradual development and long-term investment, build a new type of international relations, and build a "community of common destiny" in the field of AI to properly deal with the old problems and new challenges of global governance.

Index

ABOUT THE AUTHOR

KEVIN CHEN is a renowned science and technology writer and scholar. He was a visiting scholar at Columbia University, a postdoctoral scholar at Cambridge, and an invited course professor at Peking University. He has served as a special commentator and columnist for the *People's Daily*, CCTV, the China Business Network, SINA, NetEase, and many other media outlets. He has published monographs in numerous domains, including finance, science and technology, real estate, medical treatments, and industrial design. He currently lives in Hong Kong.